Medicine and Sport Science

Vol. 53

Series Editors

J. Borms Brussels
M. Hebbelinck Brussels
A.P. Hills Brisbane
T. Noakes Cape Town

Thermoregulation and Human Performance

Physiological and Biological Aspects

Volume Editor

Frank E. Marino Bathurst

28 figures, and 1 table, 2008

Basel · Freiburg · Paris · London · New York · Bangalore · Bangkok · Shanghai · Singapore · Tokyo · Sydney

Medicine and Sport Science

Founder and Editor from 1969 to 1984: E. Jokl[†], Lexington, Ky.

Frank E. Marino, PhD

School of Human Movement Studies and
Exercise and Sports Science Laboratories
Charles Sturt University
Bathurst, NSW 2795 (Australia)

Library of Congress Cataloging-in-Publication Data

Thermoregulation and human performance : physiological and biological aspects /
volume editor, Frank E. Marino.
p. ; cm. -- (Medicine and sport science, ISSN 0254-5020 ; v. 53)
Includes bibliographical references and indexes.
ISBN 978-3-8055-8648-1 (hard cover : alk. paper)
1. Body temperature--Regulation. 2. Exercise--Physiological aspects. 3.
Heat--Physiological effect. I. Marino, Frank E. II. Series.
[DNLM: 1. Body Temperature Regulation--physiology. 2. Athletic
Performance--physiology. 3. Exercise--physiology. 4. Heat Stress
Disorders--physiopathology. W1 ME649Q v.53 2008 / QT 165 T4112 2008]
QP135.T482 2008
612′.01426--dc22
2008031568

Contents

Preface

There are many excellent books which examine the relationship between thermoregulation and human performance. The collection of papers in this present book is in no way meant to replace those texts but rather add to the stock of knowledge in this area of human physiology. Over the past century and in particular the last decade, our understanding of the relationship between thermoregulation, performance and fatigue has changed dramatically. For example, early texts on the general subject of muscular exercise gave only a cursory account of the effect of rising body temperature on exercise [1, pp. 28–31]. Nevertheless, the common thread in many seminal texts could be summarized by the statement 'The ill-effects become more marked as the temperature of the body rises, until they culminate in the disorganization of the bodily functions, which constitute heat-stroke' [1, p. 31]. However, the majority of research since the seminal work of Pembrey and Nicol [2] shifted to understanding the cardiovascular responses with increasing body temperature because it was noted that cardiovascular strain was indeed increased during times of thermal stress. The development of muscle biopsy techniques and radioactive tracer studies permitted the study of rising and decreasing body temperature on metabolism. Further development of electromyography and electrical stimulation techniques have provided insights into the role the nervous system plays in determining human performance under thermally stressful situations.

Even with the technological and methodological advances, we still do not completely understand why thermal strain induces premature fatigue other than to presume that this is to protect the organism from irreversible cellular damage. This fact alone has necessitated further analysis of previous research in addition to considering theories derived from complimentary areas of research such as evolutionary biology, anthropology and cellular and molecular biology. This book is an attempt to propose different interpretations of the research which might explain the behavioral and physiological processes which could increase our understanding of the limitations of thermoregulation and in so doing also explain the limits of human performance.

The Earth provides extreme environments with temperature differences ranging from about −80°C at the South Pole to well over 50°C in the Sahara desert. It is not surprising then that temperature has a universal effect on life and that all life processes depend very much on temperature so that the necessary chemical reactions and subsequent biological processes take place when just the right thermal environment is present. However, exercise physiologists have not, as a general rule, considered the evolutionary history of humans as a possible avenue for answers regarding the relationship between thermoregulation and human performance. An attempt has been made by Marino in the first two chapters of this book to turn back the clock and re-discover the foundations of human evolution and the forces which may have shaped our physiology together with a comparison of the thermoregulatory strategies employed by other mammals.

A novel feature of this book is the inclusion of a 'new understanding' of exercise performance which claims that humans, like other mammals, are able to predict the requirements for successful performance in order to make use of the limited physiological capacity. Perhaps the attractiveness of this new school of thought is related to its ability to explain many observations under many different conditions. Tucker, in his paper, provides evidence for an anticipatory exercise response particularly when there is impending thermal strain. The centerpiece of this chapter is the consideration of the self-paced exercise modality which, unlike the fixed-intensity exercise modality, does not mask the individual's physiological response.

The neuromuscular system provides an exciting avenue to explore the effects of thermal strain on human performance. The last decade of research has shown that the central nervous system is deeply affected by increasing body temperature and that there is a distinct reduction in muscle recruitment when this occurs. However, the reasons for this remain elusive. Cheung provides a detailed account of the research in this area and the methodologies by which these observations have been made. Another novel area of research is that of intestinal barrier dysfunction and endotoxemia during times of increasing thermal strain. Lambert proposes that the symptoms associated with this response could possibly provide the organism with a warning signal of impending cellular disruption.

A performance enhancement method which has become increasingly popular amongst athletes is that of body cooling before exercise in order to reduce thermal strain but more recently, the application of cooling for recovery from strenuous physical exercise has been utilized. However, the effects of cooling on local muscle characteristics are seldom considered when these cooling methods are employed. Drinkwater reviews the physiological responses and the various methods utilized for cooling which could be detrimental to certain modes of exercise. Duffield, in a complimentary paper, provides an analysis of the possible benefits of cooling for thermally stressful conditions in addition to the use of body cooling for protection and recovery from physical exercise.

The morphology of humans is intimately tied to the climate in which various ethnic groups live. There is a notable difference between the athletic performances of certain ethnic groups in given events and therefore the assumption that there is a genetic advantage. Lambert and colleagues provide an in-depth analysis of the various physical characteristics of ethnic groups and the advantages that these characteristics bring in certain environmental conditions. The debate between phenotypic versus genotypic variation is explored.

Finally, the effects of thermal strain on metabolism are considered in the chapter by Mündel. However, this paper considers the consequences of metabolism and the ability for the brain to deal with these consequences rather than concentrate solely on the skeletal muscle requirements during exercise and heat stress.

I would like to acknowledge the very enthusiastic assistance of S. Karger Publishers in the preparation of this edited volume. The merit of this book lies in the work of the authors and I am indebted to them for their efforts and cooperation. Last but not least, my sincere gratitude goes to Professor Tim Noakes for his initial ideas and faith in the Editor!

Frank E. Marino

References

1 Bainbridge FA: The Physiology of Muscular Exercise, ed 3. New York, Longmans, Green and Co., 1931.

2 Pembrey MS, Nicol BA: Observations upon the deep and surface temperature of the human body. J Physiol (Lond) 1898;23:386–406.

Marino FE (ed): Thermoregulation and Human Performance. Physiological and Biological Aspects.
Med Sport Sci. Basel, Karger, 2008, vol 53, pp 1–13

The Evolutionary Basis of Thermoregulation and Exercise Performance

Frank E. Marino

School of Human Movement Studies and Exercise and Sports Science Laboratories,
Charles Sturt University, Bathurst, NSW, Australia

Abstract

The evolutionary history of mammals, but more specifically humans, indicates that heat stress was a decisive and powerful selection pressure. There is good evidence that early hominids had to adapt to a changing environment by assuming an upright posture and consequently bipedalism. Because of further distances between food sources and the need for prolonged hunting, bipedal locomotion over longer distances required higher aerobic capacities and as a consequence an increase in endogenous heat production. A cooling mechanism to balance heat loads was essential for survival and adaptations by other bodily systems such as the brain must have developed to deal with the increased heat stress. This chapter discusses the evolutionary forces which are thought to have produced the thermoregulatory system used by modern day humans in exercise performance. A particular feature that has been overlooked by thermal physiologists is the way in which mammals use the thermoregulatory system to anticipate thermal limits during physical activity and thus avoid physiological catastrophe.

The past century has witnessed extensive research into thermoregulation during exercise performance. The ability to thermoregulate effectively to avoid lethal disruption to homeostasis has been an important aspect of the research conducted, in doing so there is a greater understanding of the limits to performance under extreme conditions [1, 2]. However, this research highlights the susceptibility of living organisms to extreme environments and the practical limitations for living and performing in such conditions.

Perhaps the pinnacle in the understanding of thermoregulation was the invention of the thermometer by Galileo Galilei c1593 and its subsequent improvements culminating in the publication of *Ars de Medicina Statica* (1624) by Santorio Santorio [3]. In this publication, descriptions of temperature measurements in man were available for comparison to ill health, for at this time it was not yet known that humans regulated their body temperature around a 'set point'. What was clear, however, was that body temperature was a critical part of human physiology especially when fever was present.

The relationship between thermoregulation and human performance is intimately connected to the evolutionary forces which are seldom considered in the exercise sciences but could provide pivotal insights into the way in which modern humans deal with exercise heat stress. This chapter will discuss and outline the key evolutionary forces which are thought to have fashioned the development of the thermoregulatory system in humans with a view to providing an understanding of the limitations within which this system functions and perhaps help explain more recent findings in exercise and human performance.

The Primordial Soup and the Thermal Environment

Although there is no absolute determination of how life on Earth began, there is consensus that early life was subjected to a hot environment. The hypothesis that life began in a primordial soup was first postulated in the 1920s independently by both Oparin and Haldane [4]. In this model, the Earth's atmosphere was composed of nitrogen, ammonia, methane and hydrogen whereby heat produced chemical reactions giving rise to molecules that eventually made their way into water forming a primordial soup of amino acids and thus the building blocks of life [5]. Although the primordial soup hypothesis is not regarded as the only process by which life evolved, the point is that heat must have been an essential ingredient in this process.

Due to the sensitivity to fluctuations in environmental temperature, it would be reasonable to conclude that behavioural motivation was related to the inherent need to control the internal temperature. From this point of view it was likely that primitive organisms were able to perform metabolic functions, be excitable, reproduce and find ways of moving from one point to another. A graduation from unicellular to multicellular organisms was most likely a fundamental point in evolution and thus cellular reactions must have occurred [6]. There is very little doubt that a pivotal feature in evolution must have been the ability of organisms to respond to environmental challenges, which could only occur if structures for cellular responsiveness were available. In addition to these molecular and cellular adaptations, a major advancement in the evolution of hominids was the appearance of mammals [7]. The sequence of events leading to this is beyond the scope of this chapter and, therefore, the reader is referred to a more in-depth discussion of the events in this timeline [8].

Suffice it to say that the appearance of mammals is arguably the most important step in the evolutionary history of homo sapiens, clearly the development of upright posture and bipedal locomotion was a significant shaping force for our species [9, 10]. The relationship between upright posture and the ability to effectively regulate body temperature is not readily apparent. However, because bipedal locomotion permitted hominids to pursue prey over long distances and possibly for days [11], the production of endogenous heat must have been a significant impediment. More modern studies show that the cost of running for humans is relatively high leading to endogenous heat production for a given body mass and thus a critical factor to deal with during exercise of long duration [12, 13]. To deal with this inefficiency the development of a heat dissipating mechanism would be required for continued activity.

Evidence of Preferred Temperature of Organisms

The relationship between the origins of life and temperature were examined in a classical experiment by Mendelssohn [14] who showed that paramecia reacted to varying environmental temperatures by either dispersing or congregating according to their seemingly preferred temperature medium. When the temperature of their medium was 19°C the paramecia randomly dispersed, whereas, when the temperature was increased to 38°C they would congregate at the cooler end of 25–26°C as depicted in figure 1. What is even more remarkable is that the paramecia had a 'preferred' temperature of 24–28°C where they would congregate and avoid the extremes of either 12 or 36°C (fig. 2). Notably, the paramecia were attracted toward the moderate rather than the warmer medium suggesting that optimal function was either linked to temperature or that these unicellular organisms were capable of controlling their internal temperature by behavioural means.

Although, the temperature on much of the Earth's surface ranges from 0–50°C, some animals such as polar fish and invertebrates can live in temperatures below 0°C, whilst some algae can survive in temperatures above 70°C [15]. There is no clear evidence as to the ambient temperature in which early hominids evolved albeit that technology has permitted the ambient temperature range in which humans can live to be extended considerably. The temperature range in which humans might have evolved is an important consideration as it could explain why most animals (mammals and birds) have a set core temperature of about 37°C.

The major theories about the significance of 37°C have been constructed around the observation that death ensues in some animals following high thermal strain. A treatise on each of these is not within the scope of this review so the reader is referred to other more pertinent texts [15]. Briefly, however, these theories include (1) denaturation of proteins and thermal coagulation, (2) thermal inactivation of enzymes, (3) inadequate oxygen supply, (4) different temperature

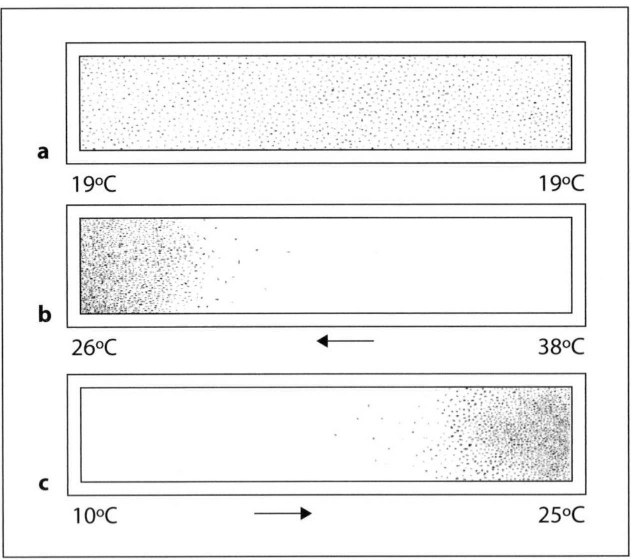

Fig. 1. Reactions of paramecia to temperature changes to their medium as studied by Mendelssohn [14]. **a** Random dispersion of the paramecia at 19°C and congregation at seemingly preferred temperature of 25–26°C by avoiding the extremes of 38°C in **b** and 10°C in **c**. With kind permission of Springer Science and Business Media.

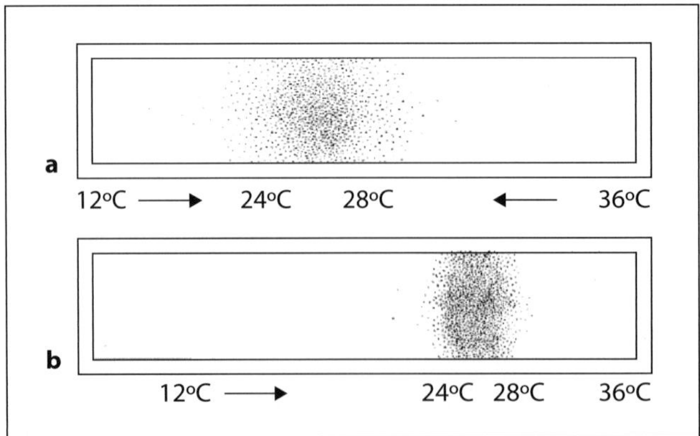

Fig. 2a, b. Congregation of paramecia at a 'preferred' temperature of 24–28°C. Note the avoidance from the extreme high (36°C) and low (12°C) temperatures and the preference to the moderate (24–28°C) temperatures [14]. With kind permission of Springer Science and Business Media.

effects on interdependent metabolic reactions know as the Q_{10} effect, and (5) temperature effects on membrane structure. Although each of these theories can partly explain the physiological processes within a narrow temperature range, they cannot explain how an organism came to 'choose' 37°C as its set temperature.

Marino

Gisolfi and Mora [16] have postulated a mathematical model based on the Law of Arrhenius and on the assumption that humans evolved and have a preference for ambient temperatures of around 25°C. The Law of Aerrhenius states that for an increase in 10°C there is an increase in the rate of chemical reactions leading to increased heat production by about ~2.3 times. Taking this law and extending it to homeotherms with the assumption that 25°C is the optimal environmental temperature for the existence of most animals (as suggested by Mendelssohn [14]), we can arrive at 37°C as an ideal core temperature like so: The temperature gradient between the core (T_c) and the environment is represented by $T_c - 25$°C. A rise in T_c of 1°C would make this equation $T_c + 1$°C $- 25$°C or $T_c - 24$°C. Heat loss will proceed according to the temperature gradient which can be expressed as $(T_c - 24)/(T_c - 25) = 1.086$, where 1.086 represents the Q_{10} effect so that a 1°C increase in T_c will increase the heat production by $2.3^{1/10}$ (or 1.086). Thus, solving for T_c results in ~37°C. Although this is a simplistic representation of how we might arrive at 37°C, it does show that 37°C might be a temperature at which heat loss and heat gain mechanisms can achieve equilibrium and that the ideal environment for most mammals could be around 25°C. This mathematical model also explains why it is important for core temperature to normally be higher than environmental temperature so that the transfer of heat is always away from the core. Conversely, if the set core temperature was lower, it may be that the sweating response would be activated at a lower threshold leading to a greater reliance on water. However simplistic this mathematical model may seem, it does connect the hypothesised evolutionary ambient temperature of ~25°C to the mammalian core temperature set point.

Other theories suggest that the thermodynamic properties of water are such that higher rather than lower temperatures are critical for maintaining thermodynamic equilibrium and therefore the least amount of stress [17]. It is also possible that thermoregulatory activities are closely related to the ratio of metabolic rate and heat conductance, so a higher set point would be essential for this relationship to hold and produce effective thermoregulation [17]. More likely, however, is that a universal thermodynamic switch exists [18]. In this model, there is a balance between entropy and enthalpy where net biological processes such as protein folding, protein-membrane interactions and protein self-assembly are possible. Any deviation from this value or that the thermodynamic switch is unable to drive the behaviour patterns related to equilibrium will most likely lead to cellular catastrophe.

Thermoregulatory Adaptations to Bipedalism

Humans have well developed sweat glands which assist greatly in sweat production and, hence evaporative cooling as shown over 50 years ago [19]. In this classic study sweat production was increased by at least 10% with acclimation and as a consequence able to account for 75–90% in body temperature attenuation when men were subjected to 1 h of physical work in a hot, dry environment. However, we now understand that the effectiveness of sweating is limited by the relative humidity by reducing the evaporative potential and cooling of the skin [20]. The ability of sweat glands to increase their effective sweat production necessitates a relatively hairless body compared with other mammals. This adaptation no doubt allows for greater conductance than would otherwise be possible with a coat of hair. Although this has not been experimentally compared in humans with other mammals, Folk et al. [21] point out that humans 'are the only living bipedal mammal with both a naked skin and a totally eccrine-dependent cooling system' [21, p. 184]. This is in contrast to those mammals that rely on apocrine glands for sweat production due to their fur coats [22]. Folk et al. [21] also point out that bipedalism necessitated a cooling system which would allow significant amounts of water to be produced so that evaporation and cooling were possible whilst hunting for extended periods; a requirement which the eccrine gland is able to sustain.

The advent of upright posture and consequent bipedalism has led to the hypothesis that the hominid brain developed and assumed its comparatively large size due to selective pressures such as gathering, offspring dependency and different eating postures becoming essential for survival and in particular, the freeing of the upper limb and hands permitting such activities as toolmaking and usage which are regarded as pivotal events in the development of the hominid brain [11]. Although these are widely held views, this has been challenged by noting that even an initial increase in brain volume cannot be accounted for by the development of rudimentary technology by virtue of the fact that increased brain size was not followed by more advanced tools and, therefore, positive feedback between these two events is not evident [23]. An alternative hypothesis proposed by Fiałkowski [24] suggests that the increase in brain size was an outcome of increased heat stress under conditions of primitive hunting. In this model, the more complex activities associated with mental tasks, communication and elaborate toolmaking could be accomplished only *after* not in advance of the increase in brain size. This hypothesis is strengthened by the proposition that early hominid environments changed from forest to grassland, having direct consequences on either the density or food type availability, the consequence of which was increased distances between forest patches and widespread increase in savannah habitat necessitating a more efficient mode of locomotion [9]. For as the accepted evolutionary theory suggests, hominids first appeared on the African continent, and therefore, were subjected to drier, warmer temperatures during the time when bipedalism is thought to have evolved [25].

Heat Stress as an Evolutionary Selection Pressure

The ability to tolerate high heat loads during locomotion must have resulted in adaptations to bodily systems to deal with heat strain. For example, it is widely accepted that the organ which is protected from thermal injury during heat stress is the brain [26]. This view has probably gained acceptance as it is thought that some mammals are able to selectively cool the brain during times of heat stress and by extension neuroprotection is assumed to be critical amongst all mammals. Increased heat stress as a selection pressure directly influencing hominid brain development is an attractive hypothesis as it strengthens the role that the brain is thought to have in protecting the organism from thermal injury. What is not clear is how a larger brain might assist in maintaining thermo-homeostasis. To answer this question, Fiałkowski [24, 27] has made use of a mathematical model which proposes that 'the complete system must be organised in such a manner that a malfunction of the whole automaton cannot be caused by the malfunctioning of a single component …but only by the malfunctioning of a large number of them' [24, p. 289]. In the present case, a larger brain has a greater chance of having many functioning parts so damage to one part may not necessarily lead to damage in another due to the number of complex interconnections. Others have also made the point that central thermoregulation is highly redundant as there are numerous brainstem regions which function in parallel in the integration of thermal inputs thereby providing a hierarchical organisation suggesting that thermoregulation in mammals is unlikely to have evolved as an independent system [28]. That is, the thermoregulatory system in mammals is highly dependant on the other systems such as the cardiovascular, integumentary and respiratory systems for integration of effector responses. The point is that, in this model the evolution of a larger brain in hominids in addition to the use of existing systems has effectively built-in a reliability/redundancy component whereby thermal injury can be either minimised or abated rather than be widespread to the whole organism thus avoiding catastrophic failure.

However, any model which affords protection to the brain from thermal injury must by necessity also explain why an organ which is thought to be pivotal in controlling thermohomeostasis should be so sensitive to changes in temperature. There are no available data indicating that brain tissue is more or less sensitive to thermal injury compared to other tissue. A threshold for irreversible damage to brain tissue due to thermal strain has never been established at least in humans, although almost a century ago it was shown that irritability of frog nervous tissue decreased as temperature increased from 36.5–44°C (fig. 3) [29]. This is a salient point as neuroprotection from thermal injury must be in excess of the damage occuring in other tissues at temperatures of 42–45°C [30] as it seems counterproductive to have the brain more sensitive to heat than other tissues or there is some

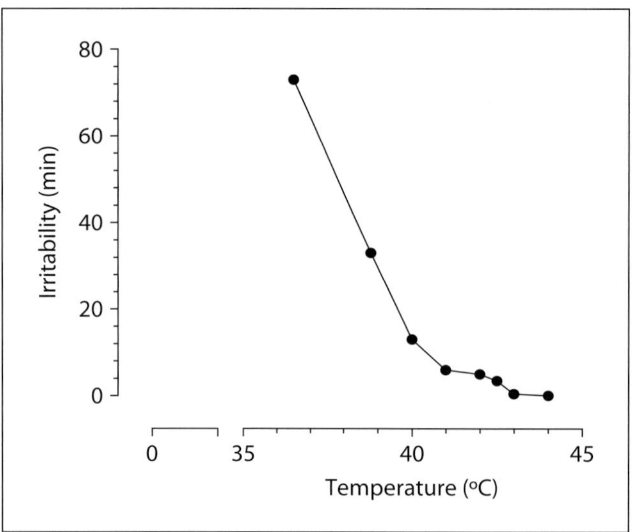

Fig. 3. Irritability of frog nervous tissue as measured by the amount of time that the nerve was able to be stimulated by electrical current and cause muscle contraction at increasing temperature. Note the reduction in irritability at about 75 min at 36.5°C to only a few seconds at about 44°C. Data are redrawn form Halliburton [29].

form of protection against such injury. Given that selective brain cooling during heat stress has never been adequately proven in humans and even doubtful in other mammals [31–33], it is plausible that protection from thermal injury to the brain might be explained by the mathematical/reliability model proposed by Fiałkowski [24, 27]. Built-in redundancy in a biological system vital for survival is likely to be more robust and able to cope with a changing environment and a plausible step in the evolutionary process.

Bipedalism and Teleo-Anticipation

An important feature of bipedalism is the capacity to run for long periods and as such humans are regarded as exceptional endurance runners compared to other mammals. In fact, there is evidence in several existing 'primitive' cultures across the Americas and Africa that humans are able to run their prey 'into the ground'; in some instances for up to two days when the prey would drop from exhaustion rendering it so much easier to slay [34, 35]. In addition to hunting, the effectiveness and ability to run long distances was very likely also exploited by primitive and ancient civilizations to carry messages and ceremonial objects for distances of up to 500 miles (~804 km) [36]. No doubt this capacity to run for long periods is contingent upon an aerobically adaptive energy system coupled to a superior cooling system in addition to the ability of primitive humans to be able to plan and subsequently pace their strategy to exploit these physiological systems to their capacities.

In essence, the ability to anticipate impending physiological limits is an essential survival component for any organism and can be identified not only in mammals but across a range of species. Teleo-anticipation was proposed by Ulmer [37] as a way of describing the relationship between perception of effort and the physiological changes that occur during heavy muscular work. In this model the organism relies on dynamic and continuously altering signals to subconsciously make adjustments to their pacing strategy in anticipation of the end point. Historically, this is not a new concept and can be traced back to the landmark writings of Mosso [38] who attempted to describe the phenomenon when he noted that quails flying from Africa to Rome needed a strategy to make the long distance flight. His subsequent studies on migrating pigeons led him to write that 'Pigeons do not become good carriers until they have been educated. Not until their third year of their education do they attain their maximum strength and skill, and the greatest perfection in their sense of direction' [38, pp. 7, 8]. A contemporary interpretation of these writings is that experience and a developed sensory feedback and feedforward calculations are an essential component of physiological systems. The ability to anticipate the required effort and make accurate calculations regarding; for instance thermal strain, must be made in a feedforward manner with sensory signals used as feedback which would be generated by a number of organs and bodily systems. The purpose of such a system would be to possibly reduce physiological catastrophe well ahead of time for as Mosso [38] noted of the quails after a long journey that: 'It is said that quails sometimes allow themselves to be caught by hand, but I cannot affirm this from personal experience, nor can any sportsman whom I have consulted' [38, p. 1]. That is to say that the quails not only were able to calculate the required effort but also had enough energy to spare should they need to avoid capture. This could only be achieved by a complex intelligent system that could calculate the physiological requirements ahead of time [39].

However, this model of teleo-anticipation has not been previously used to partly explain the process by which bipedalism might have evolved and its relationship to thermoregulation. Although there is very little empirical evidence for such an anticipatory strategy being used by early hominids, there can be no other interpretation of the retrospective analysis of the available evidence when considering the environmental challenges faced by early hominids. But, it is possible to speculate that this anticipatory or teleological ability was indeed used by early hominids by considering more recent experimental evidence on endurance athletes. First, it is worth noting that under thermally stressful conditions, athletes have been known to stop exercise at a core temperature of ~39.5–40°C, which has been coined the 'critical limiting core temperature'. This hypothesis predicts that the athlete will stop exercise at this temperature as the motivation for exercise is

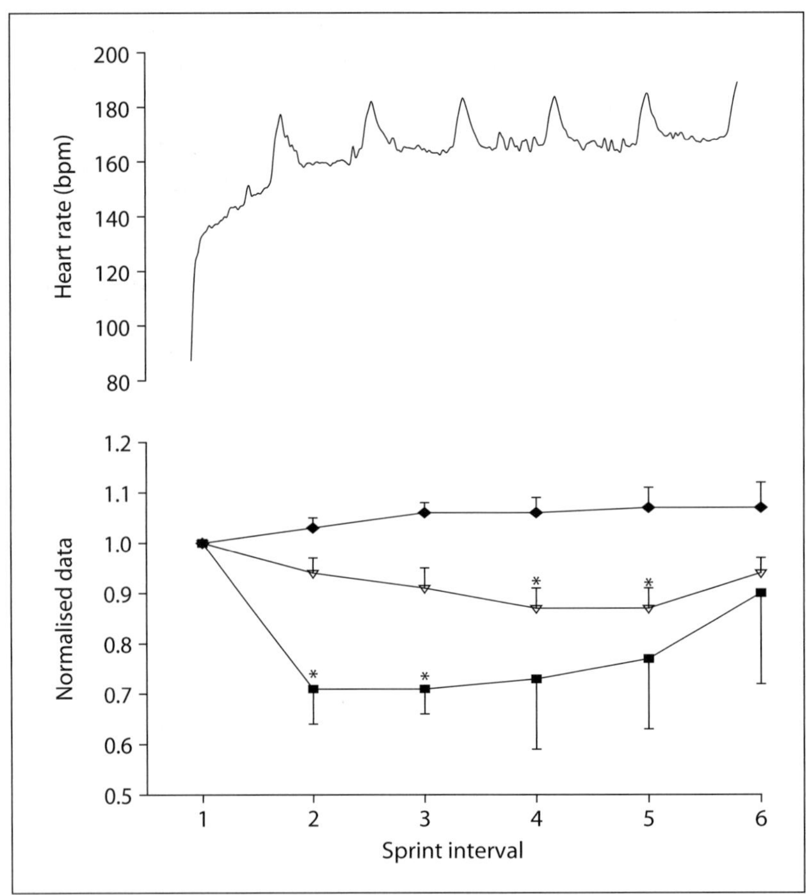

Fig. 4. Heart rate at each sprint interval (1–6) and between steady state exercise during a 60-min self-paced intermittent sprint protocol in the heat (32°C; 63% relative humidity). The heart rate (top panel) is identical at the middle of each sprint interval even though there is a decline in muscle activation as measured by integrated electromyography (IEMG; closed squares), power output (open triangles) and frequency spectrum (closed diamonds) (bottom panel) during these maximal conscious efforts. Note the return of muscle activation and power output at the penultimate and final sprints with identical heart rates [43]. *p < 0.05 compared with first sprint interval. With kind permission of Springer Science and Business Media.

significantly reduced [40]. This was later confirmed by studies which showed a coincidental reduction in neuromuscular drive from the central nervous system when body temperatures reached the critical level [41]. However, the majority of these studies have always evaluated this phenomenon with exercise at a fixed intensity rather than with more 'realistic' self-paced modalities. For the only outcome under fixed intensity modes is that the athlete always stops when the maximum thermal strain is reached. In contrast, when exercise is self-paced the athlete also stops at similar core temperatures but for all conditions, not just those in a hot environment [42; see chapter by Tucker, this vol., pp. 26–38]. Thus, the ability to adjust the rate of work is a critical factor for human performance.

One of the first studies to offer proof for the existence of such an anticipatory process showed that athletes were able to give a maximum conscious effort as evidenced by their heart rates, whilst muscle activation and power output were reduced at the same time only to be restored later in the exercise bout at identical heart rates [43] (fig. 4). These data show that the individual was regulating skeletal muscle activation in anticipation whilst maintaining a 'maximum' conscious effort so that they would be able to reach the end of the trial which would not be possible had they continued activating the same amount of muscle and generating similar amounts of heat as that in the initial stages of the trial.

A further example of such an anticipatory response under thermally stressful conditions compares Caucasian and South African distance runners [44]. This study showed that both the African and Caucasian runners were able to anticipate a thermoregulatory limit by altering their running speed in such a manner so as to avoid being unable to finish the required race distance. The Caucasian runners reduced their speed at identical rectal temperatures as the African runners in advance of the next phase of the trial, a clear indication of teleo-anticipation. For if they continued at the same speed their larger body mass would have contributed to a faster rate of heat accumulation. This evidence clearly suggests that humans are able to anticipate the thermoregulatory limits well ahead of developing hyperpyrexia. These modern studies provide experimental evidence that a teleological mechanism was perhaps a fundamental part of our evolutionary history if one accepts that heat strain was an impediment to the survival of early hominids.

Conclusions

Although there are many aspects of human evolution that have not been considered here, the relationship between thermoregulatory function and early hominid existence to the present day is indeed evident as a significant selection pressure. There is good evidence to suggest that life on Earth was heavily influenced by heat and by necessity life developed ways of minimising thermal strain. The ability of humans to develop large aerobic capacities, and therefore high endogenous heat loads necessitated an efficient cooling system which is exquisitely balanced leading to significant advantages over other animals. However, much of this physiological control is intimately reliant upon the ability to teleo-anticipate physiological limits to maintain homeostasis under a variety of conditions. It is this aspect of the evolutionary history of humans which needs further exploration.

References

1 Pandolf KB, Sawka MN, Gonzalez RR (eds): Human Performance Physiology and Environmental Medicine at Terrestrial Extremes. Dubuque, WCB, Brown & Benchmark, 1986.
2 Kerslake MD: The Stress of Hot Environments. Cambridge, Cambridge University Press, 1972.
3 Chaldecott JA: Bartolomeo Teliox and the early history of the thermometer. Ann Sci 1952;8:195–201.
4 Miller SL, Schopf WJ, Lazcano A: Oparin's 'Origin of Life': sixty years later. J Mol Evol 1997;44:351–353.
5 Miller SL: A production of amino acids under possible primitive Earth conditions. Science 1953;117: 528–529.
6 Schopf WJ: The evolution of the earliest cells. Sci Am 1978;239:111–138.
7 Valentine JW: The evolution of multicellular plants and animals. Sci Am 1978;239:141–158.
8 Kemp TS: The Origin and Evolution of Mammals. Oxford, Oxford University Press, 2005.
9 Isbell LA, Young TP: The evolution of bipedalism in hominids and reduced group size in chimpanzees: alternative responses to decreasing resource availability. J Hum Evol 1996;30:389–397.
10 McHenry HM: The pattern of evolution: studies in bipedalism, mastication, and encephalization. Annu Rev Anthropol 1982;11:151–173.
11 Krantz GS: Brain size and hunting ability in earliest man. Curr Anthropol 1968;9:450–451.
12 Marino FE, Mbambo Z, Kortekaas E, Wilson G, Lambert MI, Noakes TD, Dennis SC: Advantages of smaller body mass during distance running in warm, humid environments. Pflügers Arch 2000;441: 359–367.
13 Wright A, Marino FE, Kay D, Fanning C, Cannon J, Noakes TD: Influence of lean body mass on performance differences of male and female distance runners on warm, humid environments. Am J Phys Anthropol 2002;118:285–291.
14 Mendelssohn M: Ueber den thermotropismus einzelliger organismen. Pflügers Arch 1895;60:1–27.
15 Schmidt-Nielsen K: Animal Physiology: Adaptation and Environment, ed 4. Cambridge, Cambridge University Press, 1995.
16 Gisolfi CV, Mora F: The Hot Brain: Survival, Temperature and the Human Body. Cambridge, MIT Press, 2000.
17 McGowan C: Selection pressure for high body temperatures: implications for dinosaurs. Paleobiology 1979;5:285–295.
18 Chun P: Why does the human body maintain a constant 37-degree temperature? Thermodynamic switch controls chemical equilibrium in biological systems. Phys Scripta 2005;T118:219–222.
19 Eichna LW, Park CR, Nelson N, Horvath SM, Palmes ED: Thermal regulation during acclimatization in a hot, dry (desert type) environment. Am J Physiol 1950;163:585–597.
20 Nielsen B: Olympics in Atlanta: a fight against physics. Med Sci Sports Exerc 1996;28:665–668.
21 Folk EG Jr, Semken HA: The evolution of sweat glands. Int J Biometeorol 1991;35:180–186.
22 Robertshaw D, Talyor CR, Mazzia LM: Sweating in primates: secretion by adrenal medulla during exercise. Am J Physiol 1973;224:678–681.
23 Garn SM: Culture and the direction of human evolution. Hum Biol 1963;35:221–236.
24 Fiałkowski KR: Early hominid brain evolution and heat stress: a hypothesis. Stud Phys Anthropol 1978; 4:87–92.
25 Cerling TE, Quade J, Ambrose SH, Sikes NE: Fossil soils, grasses, and carbon isotopes from fort Ternan, Kenya: grassland or woodland. J Hum Evol 1991; 21:295–306.
26 Nybo L, Secher NH: Cerebral perturbations provoked by prolonged exercise. Prog Neurobiol 2004; 72:223–261.
27 Fiałkowski KR: A mechanism for the origin of the human brain: a hypothesis. Curr Anthropol 1986; 27:288–289.
28 Nelson D, Heat J, Prosser L: Evolution of temperature regulatory mechanisms. Am Zool 1984;24: 791–807.
29 Halliburton WD: The death temperature of nerve. Q J Exp Physiol 1915;9:193–198.
30 Jung H: A generalised concept for cell killing by heat. Radiat Res 1986;106:56–72.
31 Cabanac M: Selective brain cooling in humans: 'fancy' or fact? FASEB J 1993;7:1143–1146.
32 Fuller A, Maloney SK, Kamerman PR, Mitchell G, Mitchell D: Absence of selective brain cooling in free-ranging zebras in their natural habitat. Exp Physiol 2000;85:209–217.
33 Mitchell D, Maloney SK, Jessen C, Laburn HP, Kamerman PR, Mitchell G, Fuller A: Adapative heterothermy and selective brain cooling in arid-zone mammals. Comp Biochem Physiol [B] 2002; 131:571–585.
34 Bennett WC, Zingg RM: The Tarahumara: an Indian Tribe of Northern Mexico. Chicago, University of Chicago Press, 1935.

35 Marshall J: Man as hunter. Nat Hist 1958;72: 291–309.
36 Devine J: The versatility of human locomotion. Am Anthropol (New Series) 1985;87:550–570.
37 Ulmer HV: Concept of extracellular regulation of muscular metabolic rate during heavy exercise in humans by psychophysiological feedback. Experentia 1996;52:416–420.
38 Mosso A: Fatigue. London, George Allen & Unwin Ltd, 1915.
39 Noakes TD, St Clair Gibson A, Lambert VA: From catastrophe to complexity: a novel model of integrative central neural regulation of effort and fatigue during exercise in humans. Br J Sports Med 2004;38:511–514.
40 Nielsen B, Hales JRS, Strange S, Christensen NJ, Warberg J, Saltin B: Human circulatory and thermoregulatory adaptations with heat acclimation and exercise in a hot, dry environment. J Physiol (Lond) 1993;460:467–485.
41 Nybo L, Nielsen B: Hyperthermia and central fatigue during prolonged exercise in humans. J Appl Physiol 2001;91:1055–1060.
42 Kay D, Marino FE: Failure of fluid ingestion to improve self-paced exercise performance in moderate-to-warm humid environments. J Therm Biol 2003;28:29–34.
43 Kay D, Marino FE, Cannon J, St Clair Gibson A, Lambert MI, Noakes TD: Evidence for neuromusclur fatigue during high-intensity cycling in warm, humid conditions. Eur J Appl Physiol 2001;84:115–121.
44 Marino FE, Lambert MI, Noakes TD: Superior performance of African runners in warm humid but not in cool environmental conditions. J Appl Physiol 2004;96:124–130.

Frank E. Marino, PhD
School of Human Movement Studies and Exercise and Sports Science Laboratories
Charles Sturt University
Bathurst NSW 2795 (Australia)
Tel. +61 2 6338 4268, Fax +61 2 6338 4065, E-Mail fmarino@csu.edu.au

Marino FE (ed): Thermoregulation and Human Performance. Physiological and Biological Aspects.
Med Sport Sci. Basel, Karger, 2008, vol 53, pp 14–25

Comparative Thermoregulation and the Quest for Athletic Supremacy

Frank E. Marino

School of Human Movement Studies and Exercise and Sports Science Laboratories,
Charles Sturt University, Bathurst, NSW, Australia

Abstract

There are a number of different strategies used by animals to effectively deal with the changing environment. The various thermoregulatory strategies employed by mammals can be a critical factor determining the survival and physical performance in a range of conditions. However, it is not readily appreciated that mammals regulate their body temperature in different ways and it is usually assumed that the mechanisms for temperature regulation are very similar amongst all endotherms. In this chapter, the African hunting dog and the cheetah are used as examples of endurance and sprinting athletes, respectively to illustrate and compare the effective thermoregulatory strategies employed and to contrast the fundamental differences between these animals and humans in their approach to dealing with heat loads. Of particular interest are the strategies used by mammals to avoid developing hyperpyrexia and thereby maintain cellular homeostasis.

The ability to effectively thermoregulate can be a determining factor in human performance as multiple systems are involved in the effector response to deal with the developing heat load. Animal models have been particularly useful in the study of thermoregulation due to the wide variety of exposures that can otherwise not be undertaken with human subjects. For example, the aetiology of heat stroke and mortality has been studied in the rat model whilst reports from athletic competitions and combat conditions have provided useful insights into the reactions of humans to thermally stressful situations [1–6]. Although animal models have yielded useful physiological paradigms for studying and understanding human thermoregulation, these observations are thought to have limited applicability as human thermoregulation is not necessarily similar to that in other animals. Because the vast majority of mammals regulate body temperature around a similar set point, the general misconception is that all homeotherms regulate this set point in similar ways. Given that humans are the only hairless animals to posses eccrine sweat glands compared with

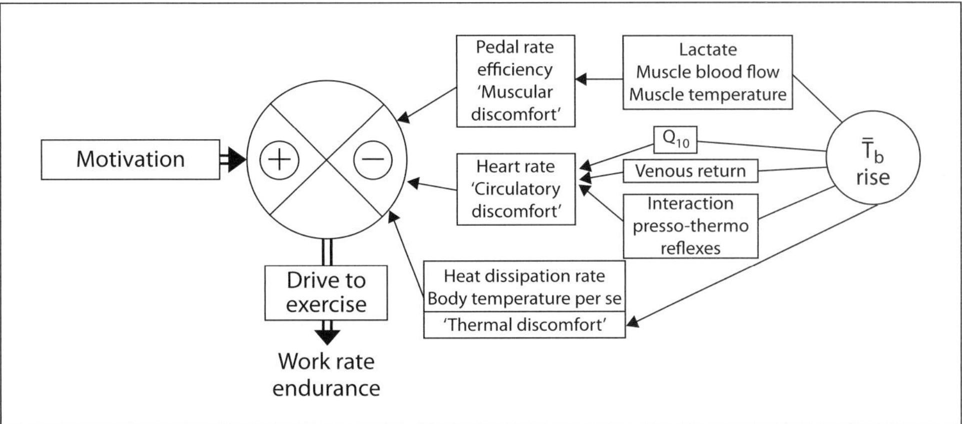

Fig. 1. Physiological factors thought to be responsible for limiting endurance exercise in the heat as proposed by Brück and Olschewski [12] (with permission). Note that the drive to exercise was influenced by either the signals generated by physiological systems and by motivation. Although not recognised at the time, motivation could be regarded as a factor related to central nervous system drive which we now understand to include down-regulated motor output to the skeletal muscle [23].

other mammals which generally have a coat of hair and posses exocrine sweat glands, indicates the operation of a very different thermoregulatory strategy [7].

These specific differences not withstanding, the means by which different species might regulate internal temperature could provide valuable insights into how heat strain induces premature fatigue which has been shown consistently among several mammals [3, 4, 8–10]. Although the relationship between exercise-hyperthermia and fatigue has been shown in both animals and humans, the causative mechanism is still somewhat elusive. Nielsen et al. [11] were likely the first to suggest the existence of a critical core temperature of around 39.5–40.0°C when exercise is terminated and coinciding with reduced muscle force production; although, Brück and Olschewski [12] were the first to hypothesise that exercise-hyperthermia lead to reduced motivation as shown in their schematic (fig. 1). In this model heat strain induces physiological responses such that all systems approach their limits and the motivation to continue exercise is diminished presumably to preserve cellular homeostasis.

Ultimately, the outcome of thermoregulation is to achieve thermal balance; however, the strategies by which this balance occurs are very different across species. For instance, the kangaroo *(Megaleia rufa)* spreads saliva for cooling at rest whilst reserving the sweating response for exercise [13]; small gazelles allow body temperature to rise above ambient temperature to conserve water [14]. This raises the question as to why there are different means of reducing heat strain in

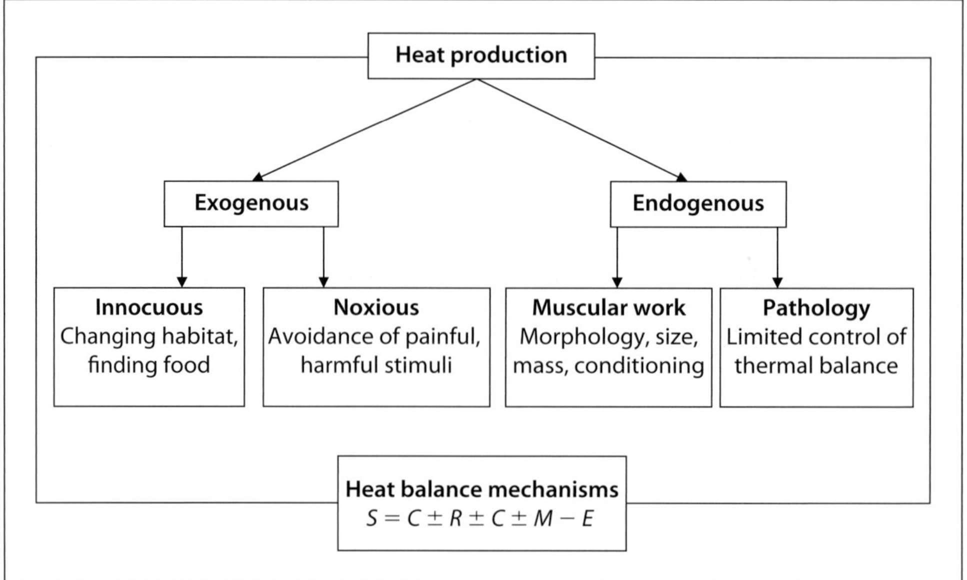

Fig. 2. Schematic drawing showing the various factors which contribute to heat production. The equation represented by $S = C \pm R \pm C \pm M - E$ is the standard heat balance equation where all avenues of heat gain and loss are represented, where S is heat storage, C is convection and conduction, M is metabolic heat, R is radiation and E is evaporation.

order to achieve a similar thermal outcome and preserve a similar set-point temperature, for if the thermoregulatory system is designed to defend the organism from thermal injury, why is it necessary to do so with a variety of strategies rather than just the one? To this end it is apparent that all animals thermoregulate in such a way so that the lethality of extreme temperature is never reached and is most likely the result of an in-built advanced (teleo-logical) warning system which reduces the prospect of developing heat illness [15].

Therefore, the purpose of this chapter is to highlight the different thermoregulatory strategies utilised by mammals to reduce thermal strain during exercise and present a more novel hypothesis which argues for an anticipatory ability to avoid a dangerous level of hyperthermia.

Avoiding Hyperpyrexia

Hyperpyrexia is commonly defined as a core temperature of 40–42°C with the possibility of serious and irreversible cellular damage [16]. Homeotherms can increase heat production in two distinct ways; either exogenously or endogenously (fig. 2). In the former, the temperature stimuli can be either innocuous or noxious. An innocuous temperature stimulus may be related to such behaviour as finding food or changing to a more habitable environment, whereas, a noxious stimulus requires avoidance strategies to minimise the pain and limit the exposure to harmful situations. Humans have dealt with innocuous situations through the use of technology and shelter, but our ability to minimise noxious stimuli still requires rudimentary survival mechanisms such as reflexes. Endogenous heat production can be achieved by either muscular exercise or shivering or due to pathological conditions such as malignant hyperthermia or multiple sclerosis [17, 18].

During physical exercise it is the combination of exogenous and endogenous stimuli that provide the thermoregulatory challenge which subsequently limits performance [19]. However, the fact that hyperthermia is a gradual and evolving condition and not normally recognised until it occurs makes it difficult to predict. It is this gradual increase in body temperature that can lead to heat illness or death should the drive for exercise not be kept in check. The fact that hyperthermia is an evolving condition during exercise is not readily appreciated for the overriding paradigm which drives the research in this area is related to the existence of a critical limiting core temperature rather than to a continuous thermal signal. If one accepts that hyperthermia is an evolving condition then it follows that heat illness can either be avoided or prevented which is indeed the case in the vast majority of circumstances. The more recently suggested paradigm explaining the avoidance of hyperpyrexia is the apparent reduction in muscular drive and the ability to anticipate critical levels of thermal strain to attenuate further heat production, a strategy which seems to occur from the outset of exercise [20, 21].

One of the most salient differences between humans and other mammals is that we will purposefully undertake physical training in a range of environments to invoke physiological adaptations so that subsequent exercise and exposure in the adaptive environment becomes less demanding, although the adaptations only last for as long as the stimulus is present, after which decay over time is expected [22]. In this regard, other mammals, unless they are specifically bred for competition or sporting events (race horses, etc.), do not undertake training per se for further 'fine tuning' and, therefore, in their natural habitat rely on intact physiological systems and natural selection for survival. However, this should not be mistaken to imply that other mammals, are better equipped to handle the environmental extremes compared to their human counterparts, but it does highlight the point that physical training induces physiological adaptations without necessarily altering physiological limits but rather training allows for a wider margin of response before those limits are reached [11]. This effect was shown in humans who were acclimated to the heat and were able to exercise for a longer period compared to when they were not acclimated; however, the thermoregulatory limit as measured by core temperature remained unchanged (fig. 3) [11].

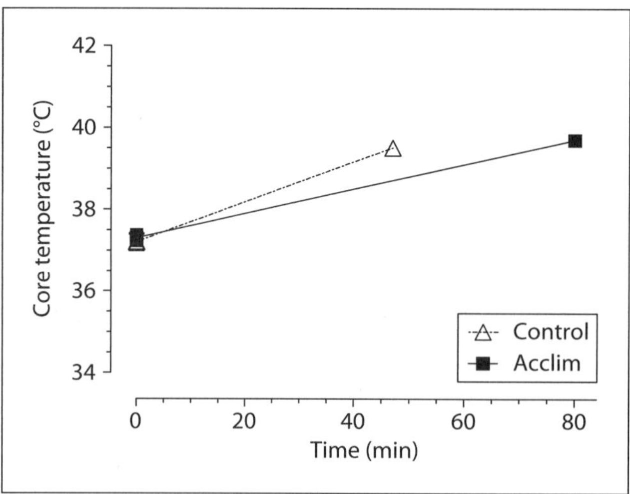

Fig. 3. These data are re-drawn from the classic study by Nielsen et al. [11]. The time to fatigue was increased due to acclimation; however, the core temperature at the termination of exercise was identical for both acclimated and control subjects. This result advanced the theory that there exists a critical limiting core temperature which reduces the motivation for continued muscular contraction.

In summary, thermoregulatory strategies among mammals vary, although there is one fundamental aspect that is seemingly identical across all mammals, that being the ability to avoid a lethal hyperpyrexia well before it can occur, the mechanism by which is still relatively unknown.

Thermoregulatory Strategies for Athletes

The modern athlete is able to expend an enormous amount of energy either over very short distances such as in sprinting or over longer distances such as the marathon. Clearly, the physiological systems which are able to deliver this energy are exquisitely balanced as athletes seldom exceed the capacity of these systems. The thermoregulatory system is usually not identified as a limiting factor for humans during short duration events but is typically thought to be a major limitation in endurance events especially in the heat [19]. However, this concept is based entirely on the assumption that heat accumulation is a precursor to fatigue which coincides with the critical limiting temperature as discussed earlier [11, 23]. This paradigm also relies on the observation that body temperature during short intense exercise does not rise sufficiently to cause fatigue presumably because the energy contribution is predominantly glycolytic and that the classical understanding is that energy usage during short intense efforts outstrips energy supply [24, pp. 31–42]. This, as we shall see is not transferable to all mammals as the capacity to undertake either intense short or long duration exercise is inextricably linked to the thermoregulatory system. For the purpose of illustration and comparison to humans, the African hunting dog *(Lyacon pictus)* and the African cheetah *(Acinonyx jubatus)* are used.

The Endurance Athlete

Studies from the 1930s show the domestic dog's *(Canis lupis familiaris)* ability for prolonged exercise to be limited by rising body temperature [25, 26]. In a classic study by Young et al. [27] the domestic Beagle whilst running at a constant speed increased its rectal temperature according to the inclination of the treadmill. The interesting finding in this study was that skin and fur temperatures were relatively stable until a gradient of 12°C, after which fur and skin temperature increased markedly suggesting that in these domestic animals heat loss at lower gradients was primarily achieved via the respiratory tract. A further observation was that '...temperature limits performance through a direct effect on the body ...and that running time would be related to the quantity of heat stored in the body' [27, p. 842]. This is a critical finding given that this study was published in 1959 and that these authors speculated that '...deterioration in ability to perform hard work in the dog is due to the same factors which limit performance in man in a hot environment' [27, p. 843]. The factors which might limit humans were not identified by the authors; however, what is most interesting about this conclusion is that it was not until 1987 that these limiting factors in humans were outlined [12] (fig. 1).

The African hunting dog, a distant member of the *Canidae* family is recognised as a relentless hunter. The parallel between humans and the African hunting dog is evident as endurance is a requirement for long distance runners and for the hominid hunters in our recent history [see chapter by Marino, this vol., pp. 1–13]. The only available study on the exercising African hunting dog shows this animal is able to run at a higher rectal temperature compared to the domestic dog [28]. Figure 4a shows that the African hunting dog can increase its rectal temperature more rapidly and to a higher value than the domestic dog in identical ambient conditions (rectal temperatures of ~41.2°C and 39.2°C, respectively). Perhaps more salient is that figure 4b shows that the African hunting dog is able to increase its rectal temperature at rest above that of the ambient temperature whilst the domestic dog maintains a constant core temperature in identical ambient conditions.

A higher core temperature during both rest and exercise seems counter-intuitive as a means of protecting the organism from the consequences of hyperthermia as a cooler core would be the preferred response during endurance performance if a maximal core temperature was the limiting factor. Interestingly, the African hunting dog loses heat through non-evaporative means, whereas the domestic dog loses most of

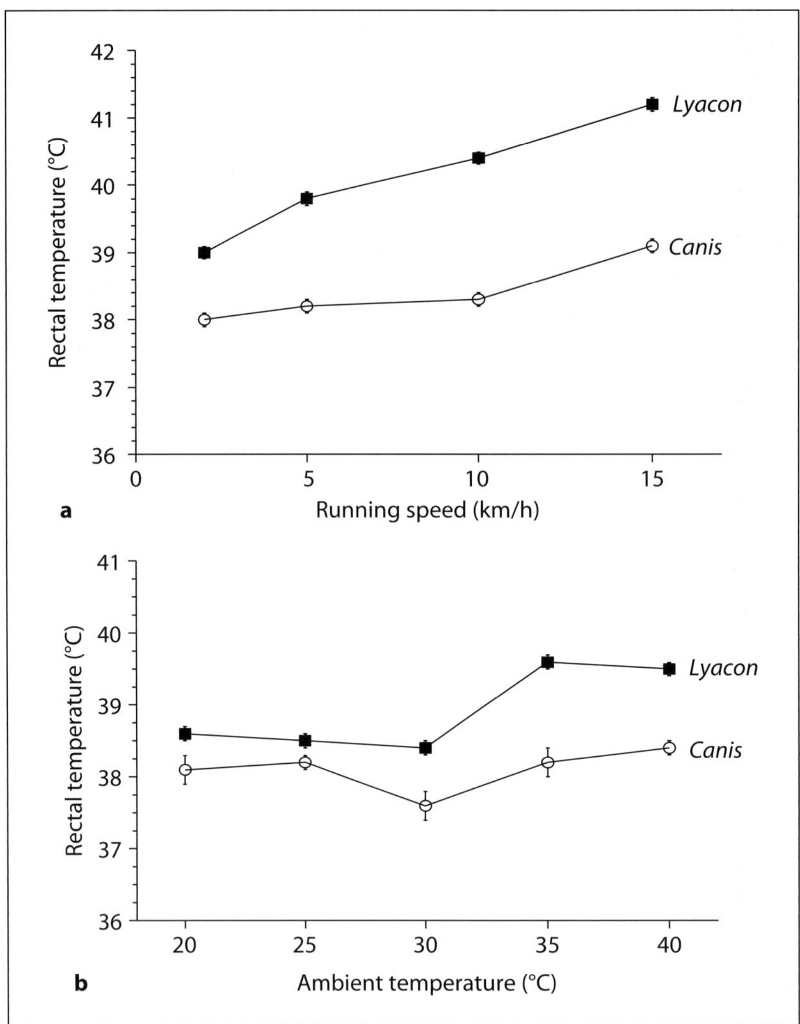

Fig. 4. a Increase in rectal temperature of the African hunting dog *(Lyacon)* and the domestic dog *(Canis)* with increasing running speed in 26°C and 25% relative humidity. **b** Increase in rectal temperature at rest with increasing ambient temperature for the same animals. Data are redrawn from Taylor et al. [28].

its heat through respiratory evaporation [28]. Therefore, the African hunting dog runs and sustains a higher body temperature without invoking a high sweat rate, thus conserving water during a long chase. In contrast, the domestic dog and likewise humans, increase body temperature according to the intensity of exercise independent of ambient temperature, although the rate of increase is greater in warmer conditions [19]. Unfortunately for the human athlete, evolution has not conferred the innate ability to increase core temperature above the environmental temperature in order to conserve water. Additionally, we are limited in heat dissipation through sweating as the most proficient means of heat loss during long duration exercise and unlike the domestic dog the respiratory tract is not an avenue for significant heat loss. Therefore, conserving water as a means of increasing our endurance capacity is not within the evolutionary make up of humans as the advent of bipedalism permitted the carrying of water and hence an increased running range.

However, recent evidence now indicates that the human athlete is equipped with an advanced warning system which is invoked under certain conditions and most evident during exercise-heat stress. The study by Marino et al. [29] showed that athletes were able to anticipate the rate of increase in body temperature and adjust their running speed accordingly to avoid developing hyperpyrexia. In this study the smaller African runners were able to maintain their running speed above that of the heavier Caucasian runners even though both sets of runners commenced running at identical core temperatures. Clearly, these data indicate that the larger Caucasian athletes could not match the running speed of the African runners if they wanted to finish the set performance run; for a higher running speed would have resulted in a higher rate of increase in core temperature and an earlier termination of exercise [30]. This model of anticipatory regulation has been shown in a number of other studies and has been explained as a subconscious feed-forward mechanism directly altering the number of motor units recruited at a given time during exercise [20, 21, 31].

In summary, the human endurance athlete is limited in the thermoregulatory strategies available to maximise performance regardless of acclimation status. Conversely, the African hunting dog is equipped with the innate ability to reduce water loss by raising body temperature above the ambient temperature even in hot conditions, whereas the human athlete has developed an innate ability to use an advanced warning system which until recently has not been recognised as a physiological event [32].

The Explosive Athlete

There are many animals that are considered fast including the horse, antelope, hare and gazelles but the cheetah is considered to be the fastest animal over a short distance [33]. The cheetah is thought to be able to run at speeds up to 114 km/h whereas the next fastest animal, the antelope can manage a maximum speed of ~80 km/h [34]. The best human sprinters can achieve a maximum speed of about 41–43 km/h over a 100 m sprint which is almost three times slower than the cheetah [35]. Not unlike the cheetah, humans can only maintain maximum speed for distances up to 150 m compared to the cheetah's 300 m. Humans can only maintain these speeds for short distances because the energy requirements of such a short dash are utilised well before the replenishment of energy can take place and presumably this is the case for the faster cursorial animals. However, human sprinters can recover from such an explosive effort and repeat the sprint only after a short rest period as shown in the studies examining the repetitive sprint ability of athletes [36]. However, this is very much dependant on whether the repeated sprints are undertaken in warm or cool conditions over an extended period of time (15–40 min) [37]. In contrast, the cheetah's extraordinary sprinting performance induces fatigue and requires a seemingly long recovery period. It is the difference in recovery period which is critical in the understanding of thermoregulatory differences and how these might impact on physical performance across different species.

Histochemical studies confirm that the muscles of the cheetah are endowed with '...an extreme profile for the support of anaerobic-based metabolism that was consistent with the animal's hunting behaviour' [38, p. 532]. In this regard, Williams et al. [38] report that the wild cheetah's fast twitch fibres comprise almost 83 and 61% of the total population of the v. lateralis and gastrocnemius muscles, respectively. However, Williams et al. [38] point out that the fast twitch muscle fibre distribution of up to 70% in human sprinters compares favourably with those of the cheetah so that in each case the fibre type distribution supports the sprinting ability of both trained humans (~40 km/h) and cheetah (~110 km/h) at their respective speeds. But unlike the human sprinter the cheetah is able to generate significant amounts of heat within a short period. In the classic study by Rowntree and Taylor [39], the running cheetah was compared to the goat *(Capra hircus)* at various speeds making it evident that when the cheetah increased its speed from 2 to 11 km/h it stored up to 70% of the heat generated compared to only 34% for the goat running at 9 km/h. Most striking, however, was that at speeds above 17 km/h about 90% of the heat generated by the cheetah's effort was stored and when body temperatures reached 40.5–41°C the cheetah refused to run. Figure 5 compares the percent heat storage at increasing running speeds for the cheetah, African hunting dog and the domestic dog. There is a stark difference between the amount of heat generated and stored by the cheetah compared to other mammals and interestingly estimates for percent heat storage for human athletes is around 8% at similar speeds and environments [40].

The interesting point is that these data also confirm why the cheetah needs to produce a successful chase within the first 200 m, for the body temperature after this point will undoubtedly be close to intolerable. This thermoregulatory strategy relies on the cheetah's ability to make judgements well ahead of initiating the chase, as an unsuccessful chase results in prolonged rest periods and ultimately no food for that day. The cheetah relies on an in-built warning system that enables it to anticipate physiological limits. This aspect of the cheetah's skill is illustrated by

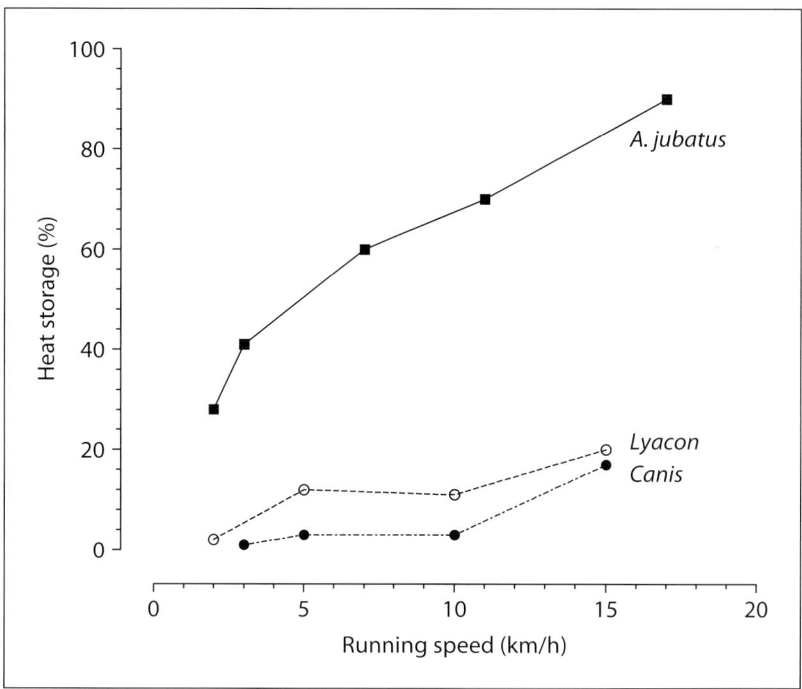

Fig. 5. Comparison of the % heat storage for the cheetah *(A. jubatus)*, African hunting dog *(Lyacon)* and domestic dog *(Canis)* at increasing running speeds. Note the significant heat storage of ~90% as the cheetah approaches 17 km/h. Data redrawn from Taylor and Rowntree [39].

observations in the wild where the adult female will release live prey for the young to practice chasing and capture [41]. Not unlike training in which humans partake, the purpose of which is to establish a template to be used for comparison and to anticipate the energy requirements and avoid cellular catastrophe [42]. But what is also critical here is that the cheetah's capacity for evaporative cooling is extremely low and it cannot rely on evaporation as a mechanism to keep its body temperature in check. This would explain why the cheetah would need a prolonged recovery period following a chase as the potential for further heat storage would be lethal.

In this sense, fatigue for the cheetah is the result of obtaining a heat storage limit. Conversely, the endurance athlete, in the present example the African hunting dog and the human cannot use heat storage as a thermoregulatory strategy as the capacity for this is limited. Rather, these mammals must rely on other mechanisms to balance the heat production either by increasing the body temperature above ambient temperature as does the African hunting dog, or invoke a cooling mechanism such as the evaporation of sweat as in the case of humans or by respiratory evaporation such as the domestic dog. In each instance, the performance of the athlete is inextricably linked to the thermoregulatory strategy available.

Conclusions

The purpose of this chapter was to differentiate the thermoregulatory strategies used by mammals to balance the thermal load during periods of either prolonged or short duration exercise. The prime examples used for comparison to humans were the African hunting dog and the cheetah. Each mammal employs a strategy that has advantages and limitations, but the overriding requirement is the balancing of endogenous heat production to avoid any possibility of developing hyperpyrexia. Regardless of the thermoregulatory strategy employed, it seems that all animals have developed the innate ability to anticipate physiological limits and reduce the likelihood of cellular disruption.

References

1 Hubbard RW, Bowers WD, Matthew WT, Curtis FC, Criss REL, Sheldon GM, Ratteree JW: Rat model of acute heatstroke mortality. J Appl Physiol 1977;42:809–816.
2 Hubbard RW, Matthew WT, Criss REL, Kelly C, Sils I, Mager M, Bowers WD, Wolfe D: Role of physical effort in the etiology of rat heatstroke injury and mortality. J Appl Physiol 1978;45:463–468.
3 Fuller A, Carter RN, Mitchell D: Brain and abdominal temperatures at fatigue in rats exercising in the heat. J Appl Physiol 1993;84:877–883.
4 Walters TJ, Ryan KL, Tate LM, Mason PA: Exercise in the heat is limited by a critical internal temperature. J Appl Physiol 2000;89:799–806.
5 Hubbard R, Mager M, Kerstein M: Water as a tactical weapon: a doctrine for preventing heat casualties. Army Sci Conf Proc 1982:125–139.
6 Sparling PB: Editorial: Environmental conditions during the 1996 Olympic Games: a brief follow-up report. Cl J Sports Med 1997;7:159–161.
7 NcGowan C: Selection pressure for high body temperatures: implications for dinosaurs. Paleobiology 1979;5:285–295.
8 Kozlowski S, Brezezinska Z, Kruk B, Kaciumba-Uscilko H, Greenleaf JE, Nazar K: Exercise hyperthermia as a factor limiting physical performance: temperature effect on muscle metabolism. J Appl Physiol 1985;59:766–773.
9 Geor RJ, McCutcheon LJ, Ecker GL, Lindinger MI: Heat storage in horses during submaximal exercise before and after humid heat acclimation. J Appl Physiol 2000;89:2283–2293.
10 Caputa M, Feistkorn G, Jessen C: Effects of brain and trunk temperatures on exercise performance in goats. Pflügers Arch 1986;406:184–189.
11 Nielsen B, Hales JRS, Strange S, Christensen NJ, Warberg J, Saltin B: Human circulatory and thermoregulatory adaptations with heat acclimation and exercise in a hot, dry environment. J Physiol (Lond) 1993;460:467–485.
12 Brück K, Olschewski H: Body temperature related factors diminishing the drive to exercise. J Physiol Pharmacol 1986;65:1274–1280.
13 Dawson TJ, Robertshaw D, Taylor CR: Sweating in the kangaroo: a cooling mechanism during exercise, but not in the heat. Am J Physiol 1974;227:494–498.
14 Taylor CR: Strategies of temperature regulation: effect on evaporation in East African ungulates. Am J Physiol 1970;219:1131–1135.
15 Marino F: Anticipatory regulation and avoidance of catastrophe during exercise-induced hyperthermia. Comp Biochem Physiol, Part B 2004;139:561–569.
16 Harker J, Gibson P: Heat-stroke: a review of rapid cooling techniques. Intensive Crit Care Nurs 1995;11:198–202.

17 Mickelson JR, Louis CF: Malignant hyperthermia: excitation-contraction coupling, Ca^{2+} release channel, and cell Ca^{2+} regulation defects. Physiol Rev 1996;76:537–591.

18 Baker D: Multiple sclerosis and thermoregulatory dysfunction. J Appl Physiol 2002;92:1779–1780.

19 Gallloway SDR, Maughan RJ: Effects of ambient temperature on the capacity to perform prolonged cycle exercise in man. Med Sci Sports Exerc 1997; 29:1240–1249.

20 Tucker R, Rauch L, Harley YXR, Noakes TD: Impaired exercise performance in the heat associated with an anticipatory reduction in skeletal muscle recruitment. Pflügers Arch 2004;448:422–430.

21 Tucker R, Marle T, Lambert EV, Noakes TD: The rate of heat storage mediates an anticipatory reduction in exercise intensity during cycling at a fixed rating of perceived exertion. J Physiol (Lond) 2006; 574:905–915.

22 Pandolf KB, Burse RL, Goldman RF: Role of physical fitness in heat acclimation, decay and reincubation. Ergonomics 1977;20:399–408.

23 Nybo L, Nielsen B: Hyperthermia and central fatigue during prolonged exercise in humans. J Appl Physiol 2001;91:1055–1060.

24 Brooks GA, Fahey TD, Baldwin KM: Exercise Physiology: Human Bioenergetics and Its Applications, ed 4. Dubuque, McGraw-Hill, 2005.

25 Dill DB, Edwards HT, Talbott JH: Studies in muscular activity. VII. Factors limiting the capacity for work. J Physiol (Lond) 1932;77:49–62.

26 Dill DB, Bock AV, Edwards HT: Mechanisms for dissipating heat in man and dog. Am J Physiol 1933; 104:36–43.

27 Young DR, Mosher R, Erve P, Spector H: Body temperature and heat exchange during treadmill running in dogs. J Appl Physiol 1959;14:839–843.

28 Taylor CR, Schmidt-Nielsen K, Dmi'el R, Fedak M: Effect of hyperthermia on heat balance during running in the African hunting dog. Am J Physiol 1971;220:823–827.

29 Marino FE, Lambert MI, Noakes TD: Superior performance of African runners in warm humid but not in cool environmental conditions. J Appl Physiol 2004;96:124–130.

30 Dennis SC, Noakes TD: Advantages of a smaller body mass in humans when distance-running in warm, humid conditions. Eur J Appl Physiol 1999; 79:280–284.

31 Kay D, Marino FE, Cannon J, St Clair Gibson A, Lambert MI, Noakes TD: Evidence for neuromusclur fatigue during high-intensity cycling in warm, humid conditions. Eur J Appl Physiol 2001;84: 115–121.

32 Noakes TD, St Clair Gibson A: Logical limitations to the 'catostrophe' models of fatigue during exercise in humans. Br J Sports Med 2004;38:648–649.

33 Sharp NCC: Timed running speed of a cheetah (*Acynonyx jubatus*). J Zool (Lond) 1997;241: 493–494.

34 Taylor CR, Lyman CP: Heat storage in running antelopes: independence of brain and body temperatures. Am J Physiol 1972;222:114–117.

35 Summers RL: Physiology and biophysics of the 100-m sprint. News Physiol Sci 1997;12:131–136.

36 Newman MA, Tarpenning KM, Marino FE: Relationship between isokinetic knee strength, single-sprint performance, and repeated-sprint ability in football players. J Strength Cond Res 2004;18: 867–872.

37 Drust B, Rasmussen P, Mohr M, Nielsen B, Nybo L: Elevations in core and muscle temperature impairs repeated sprint performance. Acta Physiol Scand 2005;183:181–190.

38 Williams TM, Dobson GP, Mathieu-Costello O, Morsbach D, Worley MB, Phillips JA: Skeletal muscle histology and biochemistry of an elite sprinter, the African cheetah. J Comp Physiol 1997;167:527–535.

39 Taylor CR, Rowntree VJ: Temperature regulation and heat balance in running cheetahs: a strategy for sprinters? Am J Physiol 1973;224:248–251.

40 Marino FE, Mbambo Z, Kortekaas E, Wilson G, Lambert MI, Noakes TD, Dennis SC: Advantages of smaller body mass during distance running in warm, humid environments. Pflügers Arch 2000; 441:359–367.

41 Caro TM: Short-term costs and correlates of playing cheetahs. Anim Behav 1995;49:333–345.

42 Paterson S, Marino FE: Effect of deception of distance and prolonged cycling performance. Percept Mot Skills 2004;98:1017–1026.

Frank E. Marino, PhD
School of Human Movement Studies and Exercise and Sports Science Laboratories
Charles Sturt University
Bathurst NSW 2795 (Australia)
Tel. +61 2 6338 4268, Fax +61 2 6338 4065, E-Mail fmarino@csu.edu.au

Marino FE (ed): Thermoregulation and Human Performance. Physiological and Biological Aspects.
Med Sport Sci. Basel, Karger, 2008, vol 53, pp 26–38

Thermoregulation, Fatigue and Exercise Modality

Ross Tucker

MRC/UCT Research Unit of Exercise Science and Sports Medicine,
Department of Human Biology, University of Cape Town, Cape Town, South Africa

Abstract

There are a number of studies which have utilised exercise protocols where subjects are free to vary
the work rate rather than having it externally imposed as is the case in fixed intensity exercise. These
studies have demonstrated that exercise performance is regulated in advance of an excessive rise in
body temperature when exercising in the heat. The evidence from self-paced exercise studies indicates that exercise in the heat is regulated specifically to control the metabolic rate, thereby ensuring
that the exercise bout can be completed before thermal injury can occur. The findings from studies
utilising different exercise modalities (fixed intensity vs. self-paced) have produced an alternative
hypothesis explaining exercise regulation.

Environmental temperature has long been recognised as a critical factor affecting
endurance exercise performance. It is known that hot (30–40°C) conditions
markedly impair performance compared with cool (3–20°C) conditions. The biological mechanisms explaining this impairment have only recently begun to be
understood. It was initially hypothesized that an increase in the oxygen-independent
contribution to energy production in the heat, resulting from reduced skeletal
muscle blood flow secondary to reduced stroke volume and cardiac output explained
this phenomenon [1].

Recent studies have questioned this hypothesis and for well over a decade it has
been thought that a critical core body temperature of approximately 40.0°C exists
above which the central drive for exercise is reduced, resulting in volitional fatigue
[2]. That is, the attainment of a critical core temperature results in inhibition of the
motor control regions of the brain, a reduction in the central activation ratio and
the drive for exercise [3–7]. This 'central fatigue' model for exercise in the heat
may apply when the workload is fixed, but its relevance to exercise where the

athlete is free to vary the exercise work rate in response to physiological changes during exercise has been questioned.

Studies utilizing an exercise modality where the athlete is free to select the work rate have introduced an alternative model to explain the impairment of exercise performance in the heat. In this model, exercise performance is impaired in advance of an excessive rise in body temperature. Power output is thus regulated during exercise in the heat specifically to control the metabolic rate, ensuring that the exercise bout can be completed without the attainment of a critical body temperature or bodily harm [8–13].

The present chapter describes the findings of research studies where exercise workrate is fixed as well as those studies where workrate was free to vary and volitionally altered by the exercising athlete.

Implications of Exercise Modality: Fixed Workrate versus Self-Paced Exercise

While a detailed discussion of exercise testing modality lies outside the scope of this chapter, a brief discussion of the key differences between the two modalities is needed. In exercise studies where the workrate (for example, cycling power output or running speed) is fixed, the most common measure of performance is time to exhaustion. In contrast, studies utilizing a self-paced model allow the athlete to alter work rate, with the main measure of performance being the time taken to complete a given exercise task. Typically, this task may take the form of a time-trial of known distance, or a trial to perform a certain amount of work, or to cover the largest distance possible within a given time period.

A crucial consideration is whether imposing an unchanging workrate on the exercising athlete is representative of what occurs during normal exercise activity. That is, during normal physical activity, humans have the ability to slow down or speed up at any time. The often neglected fact is that the change in work output that results makes a significant difference to measured physiological variables. That is, changes in work output may be as important as the classically described changes in physiological functioning that accompany exercise in the heat. By extension, the choice of constant workload versus self-paced exercise has consequences for the interpretation of measured physiological changes during exercise in the heat, for it has been established that heat production is a function of running speed as given in equation 1 [14].

$$\text{Heat production during running} = 4 \times \text{mass} \times \text{velocity}. \tag{1}$$

Therefore, any reduction in velocity reduces the rate of heat production, which consequently influences heat loss mechanisms and overall rates of heat storage and body temperatures. If the 'end-point' or limit to exercise in the heat is the attainment of a critical body temperature [6, 7, 15] then the ability to reduce the heat production and heat storage may in fact be one of the most critical physiological responses during exercise in the heat; however, this has been overlooked by those studies which have fixed the workrate and forced the athlete to exercise to exhaustion.

Accordingly, a discussion of factors affecting exercise in the heat, as well the development of models to explain the physiology underlying impaired exercise performance in hot conditions is incomplete without the acknowledgement that the athlete's ability to self-select the work rate is a key factor influencing the thermoregulatory response to exercise. That is, just as cardiovascular and metabolic responses form part of the physiological response to an external stress (in this case, exercise in hot, humid conditions), so too the behavioural responses, mediated through altered neuromuscular drive and work rate cannot be discounted.

The following section describes the findings of studies in both categories of exercise testing in hot conditions – exercise at a constant workload and self-paced exercise trials. Each section describes the impact of the environmental strain on performance, together with the physiological model that is developed in order to understand the effect of the heat on performance.

Exercise at a Constant Workload – The Critical Core Temperature Hypothesis

Numerous studies have found that exercise time to fatigue is greatly reduced by high ambient temperatures [3, 4, 16]. In one such study [3], 8 subjects cycled to exhaustion at 70% of their maximum oxygen uptake ($VO_{2\,max}$) at four different temperatures: 4, 11, 21 and 31°C with relative humidity set at 70% for all trials. Exercise duration was longest at 11°C (93.5 ± 6.2 min), and shortest at 31°C (51.6 ± 3.7 min). A significantly higher overall rating of perceived exertion (RPE) was recorded in the hot (31°C) trial compared to the other three trials, and the core temperature was significantly greater in the 31°C condition compared to the 11°C condition.

Time to fatigue was also influenced significantly by ambient temperature in a study by Parkin et al. [4]. Eight endurance trained subjects cycled to fatigue at 70% $VO_{2\,max}$ at 3, 20 or 40°C with relative humidity at 40% for all conditions. Exercise times to fatigue were 85 ± 8, 60 ± 11, and 30 ± 3 min at 3, 20 and 40°C, respectively. Time to fatigue was therefore reduced by 65% in the 40°C condition compared to the 3°C condition. In a similarly designed study by Febbraio et al. [16], 12 subjects cycled at 70% of their $VO_{2\,max}$ for 40 min at 20°C with 20% relative humidity, and again at 40°C with 20% relative humidity. All 12 subjects were able to complete the workload at 20°C, but 5 subjects were unable to complete the required duration at 40°C. In another study [7], 14 subjects performed sub-maximal cycle exercise to exhaustion in a temperate (18°C) and a hot (40°C) environment. In the hot environment, RPE and core temperature rose throughout exercise until fatigue, which occurred after 50 ± 3 min with an RPE of 20 units (6–20 RPE scale, [17]). In the temperate condition, exercise continued for 1 h without exhausting the subjects, and the RPE remained at sub-maximal levels until the required 60 min had been completed.

It was based on these, and similar studies, that a model for a limitation to exercise in the heat was developed, as researchers began to notice that when the body temperature approached 40°C, the exercising subjects displayed symptoms that were characteristic of impaired neurological functioning. In 1987, Bruck and Olschewski [18] observed that subjects became dizzy, incoherent and unable to move their legs in a co-ordinated manner as they approached the end-point of exercise to exhaustion. They speculated that hyperthermia, or the attainment of a critically high body temperature, reduced the drive to exercise.

Subsequently, this hypothesis was supported by Nielsen et al. [19] following a study in which they showed that exhaustion in the heat was not caused by inadequate leg blood flow, altered muscle metabolism or skeletal muscle glycogen depletion. Subjects did, however, report dizziness and an inability to move their legs as they approached exhaustion in the heat, suggesting that mental functioning and the central nervous system are susceptible to high temperatures. It was surmised that at core temperatures of about 40°C, the ability of the motor centres of the brain to recruit motor units is reduced, perhaps by an effect of hyperthermia on the motivation for motor performance [19].

A similar hypothesis has been made in subsequent studies [2, 6, 20]. These studies have found that fatigue coincides with the attainment of a critical core temperature of ~40°C, regardless of the state of heat acclimation manipulation of pre-exercise body temperature and the rate of heat storage [2, 21]. Galloway and Maughan [3] found that exhaustion during cycling at an ambient temperature of 31°C occurred at a rectal temperature of 40.1 ± 0.2°C. Nielsen et al. [2] found that exercise was terminated at a core temperature of 39.8 ± 0.13°C on the first day and 39.7 ± 0.15°C on the tenth day of a heat acclimation protocol in which subjects exercised to exhaustion in the heat for 9–12 consecutive days. The improvement in performance measured in this study (subjects reached fatigue after 80 min on day 10 compared to 48 min on day 1) was not attributed to improved tolerance to high temperatures, but rather to the increased time it took for the body temperature to rise to the levels at which exercise was volitionally terminated [2].

Gonzalez-Alonso et al. [20] used water immersion to manipulate pre-exercise body temperatures to 36, 37 and 38°C. It was found that despite the different initial temperatures, exhaustion coincided with the same muscle (41°C), oesophageal (40°C) and skin (37°C) temperatures. In the three different conditions, the difference in oesophageal temperature at exhaustion between subjects was 0.1–0.5°C. In the same study, subjects wore a water-perfused jacket which allowed the rate of heat storage during exercise to be altered. As with previous studies, despite different rates of heat storage, all subjects reached exhaustion at similar oesophageal and muscle temperatures (approximately 40 and 41°C, respectively). It was concluded that the hypothalamic and internal organ temperatures would have reached similar levels at fatigue, and that the high temperatures reduced the central drive for exercise by influencing the motor control centre in the brain [20].

More recently, studies of the effects of heat stress on central drive, muscle function and brain activity support the hypothesis that skeletal muscle motor unit recruitment and central activation are decreased after body temperatures reach critical values. In one such study [6] subjects cycled at $\sim$60% $VO_{2\,max}$ for a maximum of 1 h at either 40°C (hyperthermia) or 18°C (normothermia), before performing one of three maximal voluntary contraction (MVC) protocols immediately at the end of the exercise. The first protocol consisted of 40 maximal knee extensions of two seconds each repeated every five seconds. The second protocol consisted of two min of sustained maximal isometric knee extension. The third protocol consisted of two min of sustained maximal isometric handgrip extension to determine the effect of elevated body temperature on a non-exercised muscle group. In all three protocols, electrical stimulation of the femoral nerve was superimposed on muscle contraction at various intervals to assess the degree of central nervous system activation. It was found that immediately after completion of the exercise bout, which was sufficient to raise body temperature to 38°C in the normothermic condition and 40°C in the hyperthermic condition, there was no difference in force development during the 40 MVCs, and the voluntary activation percentage was 92% for both conditions. The force output and activation percentage was, however, different during the sustained 2-min leg MVCs. It was found that force production was similar for the first 5 s of the contraction, but decreased to significantly lower levels in the hyperthermic trial than in the normothermic trial. This was associated with a significantly lower voluntary activation percentage (54% compared with 82%) as shown in figure 1. Significantly, the total force output, measured as the combined force output during MVC and electrical stimulation, was the same in both trials. This suggests that the capacity of the skeletal muscles to generate force is not affected by hyperthermia, but that impaired central activation is solely responsible for the difference in force output between trials [6].

Furthermore, the same pattern was evident in the handgrip contractions, indicating that central activation of muscle is independent of whether the muscle was active or not. The findings also suggest that muscle temperature does not play a large role in mediating the decline in central activation, because the

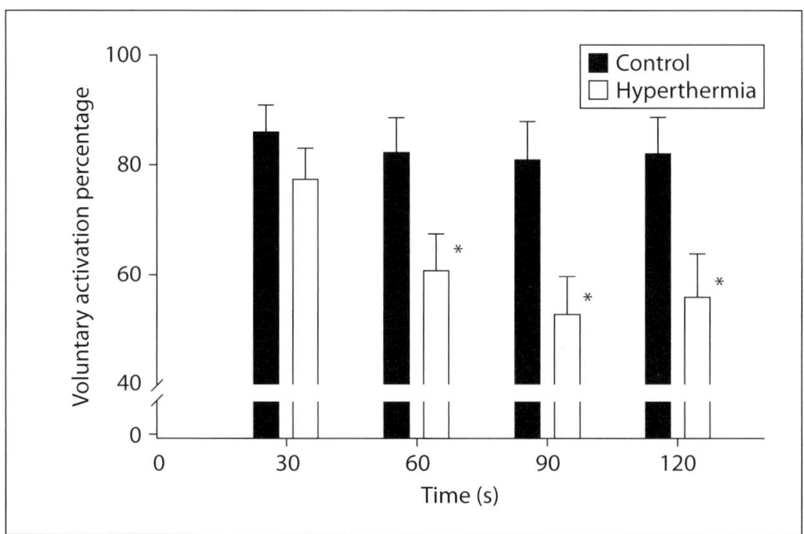

Fig. 1. Changes in voluntary activation during 2-min of sustained maximum voluntary contraction of the knee extensors during hyperthermia and control conditions. These data were generated whilst subjects undertook maximal voluntary contractions over a 2-min period during which electrical stimulation was superimposed at 30-second intervals. *Significantly lower than control condition (p < 0.05). Used with permission copyright American Physiological Society.

temperature of the active skeletal muscle was only 38.7°C for the handgrip contraction and 40.7°C for the knee MVCs. It was concluded from this study that hyperthermia causes a form of 'central fatigue', with reduced central activation leading to a lower force production [6]. That is, body temperature was elevated by the exercise bout, reaching significantly greater levels in the hyperthermic condition than the normothermic condition, and this resulted in a reduced ability of the brain to activate skeletal muscle motor units after body temperature was elevated.

Other novel studies have used electroencephalogram (EEG) measurements and cerebral blood flow experiments to show that brain function is altered during exercise in the heat [6, 7, 15] lending support to the hypothesis that central processes are affected by high core temperatures. It has been shown that alterations in the EEG of the frontal cortex are linearly related to increasing core temperatures during exercise heat stress [15]. In this particular study, the alpha/beta index was used as an indication of arousal levels, where an increase in this ratio reflects a decrease in arousal. It was found that at fatigue, the ratio was significantly greater in hot compared to cool conditions, and that increases in this ratio during exercise were significantly correlated to increases in oesophageal temperature ($r^2 = 0.98$).

Nybo and Nielsen [5] have demonstrated that subjectively perceived exertion is highly associated with increases in core temperature and with frequency changes of the EEG obtained over the prefrontal cortex. In contrast, there were no correlations between RPE and any of the measured EMG parameters. The unaltered root mean square (RMS), median frequency, and amplitude of the smoothed EMG during both the control and hyperthermic exercise trial indicate that hyperthermia did not result in fatigue-induced changes in motor unit recruitment and/or discharge rates. These results support the idea that altered activity within the central nervous system rather than changed muscular activity is involved in the development of fatigue during prolonged exercise in hot environments.

In summary, studies of exercise performance in which the workrate is fixed have revealed that a limit of approximately 40°C exists, above which volitional exercise ceases. In such studies, a fixed external work load is imposed, and the exercising athlete has no alternative but to maintain the required force output until it becomes impossible to do so, and the trial is terminated. That this volitional fatigue coincides with a body temperature of about 40°C has invited the conclusion that fatigue and hence impaired performance in the heat is the result of excessive heat storage that raises the body temperature to a critical threshold. This model, which has been termed 'catastrophic' [22], since it views fatigue as the result of a failure in one or more physiological systems (in this case, thermoregulatory and cardiovascular) fails to acknowledge that if given the choice to alter the power output, performance may in fact be regulated in advance of this critical core temperature.

Self-Paced Exercise Trials – Anticipatory Regulation of Performance in the Heat

As discussed previously, exercise in which the work rate is imposed and remains constant until fatigue is not representative of normal exercise circumstances, since athletes are able to alter their work rate in response to physiological and environmental signals or cues. The alteration in work rate, a behavioural response, is as much a part of the physiological regulation during exercise in the heat as circulatory and metabolic adjustments. However, only when exercise studies allow the athlete to self-select the pace can this important component of the homeostatic response be observed. Thus, while studies of exercise at a constant workload have contributed to understanding of physiological processes involved in exercise-induced hyperthermia, they do not appropriately reflect what is occurring during self-paced exercise. The following section describes the outcomes of studies where work rate is free to vary, and develops the model for understanding how exercise performance may be regulated in advance of thermoregulatory 'failure' [11, 22].

During self-paced exercise in the heat, performance is typically measured as time taken to complete a known distance or duration. Like constant workload trials to exhaustion, these studies have found that performance is impaired in hot environments [8, 9, 12]. However, a key distinction between the self-paced studies and the constant workload trials described previously is that self-paced trials suggest that the impairment in exercise performance is observed in advance of the critical core temperature, and often occurs despite body temperatures that are similar to those observed in cool conditions. This suggests that the 'central fatigue' model described previously is incomplete, and unable to account for observed changes in performance that occur in the absence of hyperthermia.

For example, Tatterson et al. [9] found that power output was lower in hot (33°C) than in temperate (23°C) conditions after only 15 min of a 30-min self-paced cycling time-trial, even though core temperatures rose at similar rates in the two environments. Thus, in contrast to the previously discussed model that exercise performance is impaired after body temperature reached 40°C, this finding suggests that performance is impaired before critical temperatures are reached. More significantly, the power output was reduced in the hot trial despite a core temperature that was similar to that observed in the temperature environment. It was postulated that the brain was sensitive to the rate of increase in arterial blood temperature, and selected a power output relative to the rate of rise in core temperature. The mechanism for such an adjustment in power output would presumably be a reduction in the activation of skeletal muscle motor units, as has been described previously [23, 24]. However, Tatterson et al. [9] did not measure EMG activity to confirm their conclusions.

Marino et al. [8] found that four out of 16 well-trained runners were unable to complete an 8-km time-trial following 30 min of sub-maximal running (70% of peak treadmill running speed) at an ambient temperature of 35°C with a relative humidity of 60%. Furthermore, in those runners able to complete the trial, running speed decreased progressively at 35°C, whereas at temperatures of 15 and 25°C, all runners were able to complete the trials, and running speeds were maintained through the trials, leading to significantly improved performance times in the cooler conditions.

Subsequently, Marino et al. [12] studied African and Caucasian runners in hot (35°C) and cool (15°C) conditions using the same protocol, consisting of a 30-min run at 70% of peak treadmill running speed followed by an 8 km time-trial in the two conditions. It was found that the African runners outperformed the Caucasian runners in the hot, but not in the cool condition. The performance advantage of the African runners in the heat was associated with a difference in pacing strategy between the groups. That is, the African runners began the 8 km time-trial at a faster speed than the Caucasian runners, even though the rectal temperatures of both groups were below 40°C and not different between the

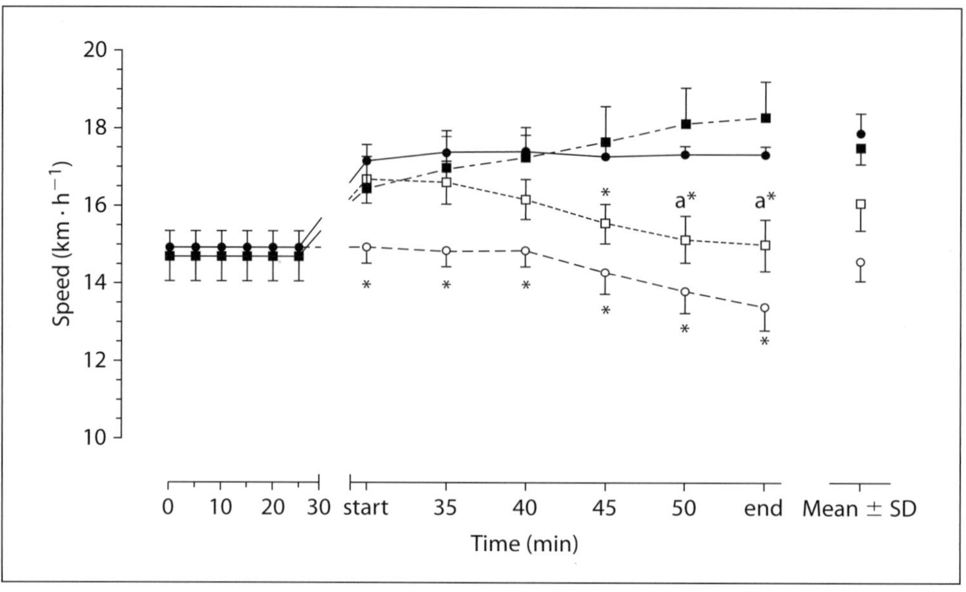

Fig. 2. Shows the running speed during a sub-maximal run for 30 min at 70% of peak treadmill velocity and during the subsequent 8 km performance run. In cool conditions (15°C, 60% relative humidity) the African runners are filled circles, Caucasian runners are filled squares. In hot conditions (35°C, 60% relative humidity) the African runners are clear circles, and Caucasian runners are clear squares. Note the lower running speed in hot conditions immediately following the 30 min run at 70% peak treadmill speed for Caucasian runners whilst the Africa runners were able to maintain speed for at least 5-min with a gradual decline. This observation would not otherwise be evident in a fixed intensity protocol. *Significant differences to cool conditions for both African and Caucasian runners; ᵃsignificant difference in hot conditions for Caucasian runners. Used with permission copyright American Physiological Society.

groups (~38.4°C) when the 8 km time-trial commenced. The running speed of the Africans was subsequently maintained at the higher levels throughout the trial, resulting in an overall improvement in performance compared to the Caucasian group as shown in figure 2. Significantly, in the cooler condition (15°C), the pacing strategy was similar between the African and Caucasian runners, suggesting that whatever factor/s caused the Caucasians to adopt a slower speed at the start of the 8 km trial in the heat were not present in the cooler condition.

This difference in pacing strategy was attributed to the smaller body size of the Africans, which would, according to equations for heat production, result in a reduced rate of heat storage at a given running speed [14]. It was suggested that the early reduction in running speed in the heat occurred due to an anticipatory response which would 'control the exercise work rate by regulating the number of motor units that are recruited or derecruited during prolonged exercise in the heat'. In the cool condition, where a limiting rise in body temperature was not a factor, the performance was similar between groups. This suggests that the pacing strategy is: (1) sensitive to the environmental conditions, the rate of heat storage and the possible rise in body temperature, and (2) a key component of the physiological response to exercise in the heat, and that under conditions where the attainment of a high, possibly limiting body temperature was possible, work rate (in this case, running speed) are reduced in anticipation of high temperatures.

Subsequently, Tucker et al. [13] showed that during 20 km cycling time-trials, power output and integrated electromyogram (iEMG) activity, an indirect measure of skeletal muscle recruitment, were reduced in hot conditions (35°C) compared to cool conditions (15°C), despite the absence of any differences in measured rectal temperature, heart rate or RPE. The reduction in efferent command thus occurred in advance of the attainment of the critical, limiting temperature of 40°C. This study was the first to provide evidence that muscle activation was reduced during dynamic exercise in hot conditions.

Significantly, power output and iEMG activity then increased significantly in the final kilometer in both conditions, when body temperatures were at their highest. Therefore, power output and iEMG activity decreased at sub-maximal body temperatures, and then increased again at the end of the trials, when body temperatures were approaching the theoretical 'limit' to exercise in the heat [6]. The 'central fatigue' theory for exercise in the heat which holds that fatigue results from a reduction in central drive caused by hyperthermia [6, 7] is clearly not appropriate, because changes in EMG activity and power output precede the attainment of this critical level or indeed any differences in body temperature compared to a cooler environment. It was concluded that an anticipatory regulatory system exists to protect the exercising athlete by mediating reductions in skeletal muscle activation and thus heat storage [13].

In a novel exercise protocol in which work rate was free to vary, but subjects were instructed to cycle at a fixed, pre-determined RPE at three different temperatures; 15°C (COOL), 25°C (NORM) and 35°C (HOT) it was found that the conscious sensation of fatigue was responsible for mediating the reduction in power output during self-paced exercise [25]. This was evidenced by the fact that the power output maintained by the cyclists decreased significantly over time in all three conditions, but that rate of decline was significantly greater in the HOT condition than in the NORM and COOL conditions. This significantly greater rate of decline in power output was associated with significantly higher rates of heat storage during the first few minutes of the trial. That is, the rate of heat storage, and not the core temperature, was the only measurable difference within the first 10 min of cycling in all three conditions. Thus, in the hot condition, the rate of heat storage in the first few minutes was significantly greater than in NORM and COOL, leading to a significant decrease in power output in HOT compared to the other two conditions. Since the reduction in power output took place while cycling at a fixed RPE, it was concluded that the RPE was sensitive to the rate of heat storage, and mediated a reduction in power output, to prevent the RPE from increasing to unacceptable levels [25].

The overall conclusions from these studies [9, 12, 13] is that alterations in pacing strategy are achieved by means of a reduced motor efferent command in response to an anticipatory calculation, perhaps based on afferent inputs regarding the rate of heat storage or skin temperature. It would appear from these studies that the direct effect of core temperature (either rectal, body or oesophageal) is relatively minor, since alterations in iEMG activity [13, 25], and work rate [9, 12, 13, 25] occur before measured body temperatures are different between conditions. Fixed work rate trials have found that high body (and brain) temperatures (>40°C) are directly responsible for fatigue, either by inhibiting activation of muscle, reducing arousal levels, or altering cerebral blood flow. However, the self-paced model presents an alternative, in that alterations in work rate may occur specifically to prevent body temperatures to rising to reach potentially harmful levels.

The specific mechanism by which this 'anticipatory' system functions is not yet completely understood. In addition to the already described hypotheses that the rate of heat storage, mediated perhaps by the elevated skin temperatures in the hot conditions, is responsible for reducing skeletal muscle motor unit activation, further studies are required to investigate brain functioning during self-paced exercise. An additional confounding variable is that exercise in the heat is associated with cardiovascular challenges, which were once thought to be the limiting factor during exercise in hot conditions. While this theory has now been proven incorrect [26], the cardiovascular challenges remain an important possible regulator of performance during time-trial performances [27]. Future studies will be required to improve measurement techniques for measuring brain function during dynamic exercise, and to investigate the effects of heat without the associated cardiovascular and metabolic changes.

Conclusion

In conclusion, both constant workload performance and self-paced exercise performance are impaired in hot compared to temperate and cool environments. The use of exercise protocols where subjects exercise to volitional exhaustion in the heat has led to the development of the theory that fatigue is the direct result of hyperthermia, where high body or brain temperatures are responsible for a failure to maintain skeletal muscle activation at the required level.

The alternative hypothesis, derived from studies where exercise is self-paced, suggests that exercise is regulated in advance of such a thermoregulatory 'failure', and that the rate of heat storage, and possibly afferent inputs from the skin are responsible for mediating changes in muscle activation and work output. The consequence of these changes is that the exercising athlete is able to complete exercise safely, though with impaired performance.

References

1 Rowell LB, Marx HJ, Bruce RA, Conn RD, Kusumi F: Reductions in cardiac output, central blood volume, and stroke volume with thermal stress in normal men during exercise. J Clin Invest 1966;45: 1801–1816.

2 Nielsen B, Hales JRS, Strange S, Christensen NJ, Warberg J, Saltin B: Human circulatory and thermoregulatory adaptations with heat acclimation and exercise in a hot, dry environment. J Physiol (Lond) 1993;460:467–485.

3 Galloway SDR, Maughan RJ: Effects of ambient temperature on the capacity to perform prolonged cycle exercise in man. Med Sci Sports Exerc 1997; 29:1240–1249.

4 Parkin JM, Carey MF, Zhao S, Febbraio MA: Effect of ambient temperature on human skeletal muscle metabolism during fatiguing submaximal exercise. J Appl Physiol 1999;86:902–908.

5 Nielsen B, Hyldig T, Bidstrup F, Gonzalez-Alonso J, Christoffersen GRJ: Brain activity and fatigue during prolonged exercise in the heat. Pflügers Arch 2001;442:41–48.

6 Nybo L, Nielsen B: Hyperthermia and central fatigue during prolonged exercise in humans. J Appl Physiol 2001;91:1055–1060.

7 Nybo L, Nielsen B: Perceived exertion is associated with an altered brain activity during exercise with progressive hyperthermia. J Appl Physiol 2001;91: 2017–2023.

8 Marino FE, Mbambo Z, Kortekaas E, Wilson G, Lambert MI, Noakes TD, Dennis SC: Advantages of smaller body mass during distance running in warm, humid environments. Pflügers Arch 2000; 441:359–367.

9 Tatterson AJ, Hahn AG, Martin DT, Febbraio MA: Effect of heat and humidity on time trial performance in Australian national team road cyclists. J Sci Med Sport 2000;3:186–193.

10 Cheung SS, McLellan TM: Heat acclimation, aerobic fitness, hydration effects on tolerance during uncompressible heat stress. J Appl Physiol 1998;84: 1731–1739.

11 Marino F: Anticipatory regulation and avoidance of catastrophe during exercise-induced hyperthermia. Comp Biochem Physiol [B] 2004;139:561–569.

12 Marino FE, Lambert MI, Noakes TD: Superior performance of African runners in warm humid but not in cool environmental conditions. J Appl Physiol 2004;96:124–130.

13 Tucker R, Rauch L, Harley YXR, Noakes TD: Impaired exercise performance in the heat associated with an anticipatory reduction in skeletal muscle recruitment. Pflügers Arch 2004;448:422–430.

14 Dennis SC, Noakes TD: Advantages of a smaller body mass in humans when distance-running in warm, humid conditions. Eur J Appl Physiol 1999; 79:280–284.

15 Nielsen B, Hyldig T, Bidstrup F, González-Alonso J, Christoffersen GRJ: Brain activity and fatigue prolonged exercise in the heat. Pflügers Arch 2001;442:41–48.

16 Febbraio MA, Snow RJ, Stathis CG, Hargreaves M, Carey MF: Effect of heat stress on muscle energy metabolism during exercise. J Appl Physiol 1994;77: 2827–2831.

17 Borg GA: Psychophysical bases of perceived exertion. Med Sci Sports Exerc 1982;14:377–381.

18 Brück K, Olschewski H: Body temperature related factors diminishing the drive to exercise. J Physiol Pharmacol 1986;65:1274–1280.

19 Nielsen B, Savard G, Richter EA, Hargreaves M, Saltin B: Muscle blood flow and muscle metabolism during exercise heat stress. J Appl Physiol 1990; 69:1040–1046.

20 González-Alonso J, Calbert JAL, Nielsen B: Metabolic and thermodynamic responses to dehydration-induced reductions in muscle blood flow in exercising humans. J Physiol (Lond) 1999;520: 577–589.

21 González-Alonso J, Teller C, Anderson SL, Jensen FB, Hyldig T, Nielsen B: Influence of body temperature on the development of fatigue during prolonged exercise in the heat. J Appl Physiol 1999; 86:1032–1039.

22 Noakes TD, St Clair Gibson A, Lambert VE: From catastrophe to complexity: a novel model of integrative central neural regulation of effort and fatigue during exercise in humans: summary and conclusions. Br J Sports Med 2005;39:120–124.

23 Kay D, Marino FE, Cannon J, St Clair Gibson A, Lambert MI, Noakes TD: Evidence for neuromusclur fatigue during high-intensity cycling in warm, humid conditions. Eur J Appl Physiol 2001;84: 115–121.

24 St Clair Gibson A, Schabort EJ, Noakes TD: Reduced neuromuscular activity and force generation during prolonged cycling. Am J Physiol Reg Integ Comp Physiol 2001;281:R187–R196.

25 Tucker R, Marle T, Lambert EV, Noakes TD: The rate of heat storage mediates an anticipatory reduction in exercise intensity during cycling at a fixed rating of perceived exertion. J Physiol (Lond) 2006; 574:905–915.

26 Savard GK, Nielsen B, Laszcynska J, Larsen BE, Saltin B: Muscle blood flow is not reduced in humans moderate exercise and heat stress. J Appl Physiol 1988;64:649–657.

27 Cheung SS, Sleivert GG: Multiple triggers for hyperthermic fatigue and exhaustion. Exerc Sport Sci Rev 2004;32:100–106.

Ross Tucker, PhD
MRC/UCT Research Unit of Exercise Science and Sports Medicine
Department of Human Biology, University of Cape Town
Cape Town (South Africa)
Tel. +27 21 650 4572, Fax +27 21 686 7530, E-Mail ross.tucker@mweb.co.za

Tucker

Marino FE (ed): Thermoregulation and Human Performance. Physiological and Biological Aspects.
Med Sport Sci. Basel, Karger, 2008, vol 53, pp 39–60

Neuromuscular Response to Exercise Heat Stress

Stephen S. Cheung

Environmental Ergonomics Laboratory, Department of Physical
Education and Kinesiology, Brock University, St. Catharines, Ont., Canada

Abstract

Hyperthermia, through either passive exposure or exercise in hot environments, can severely impair
exercise capacity, and one primary pathway of impairment may be the neuromuscular system. This is
because fatigue or reduced exercise capacity in other physiological systems ultimately terminate in
an inability of the muscle to adequately generate the required force to maintain a desired workload.
In addition, the functional capacity of the neuromuscular system, from the central activation of the
motor unit pool, through neural transmission along the peripheral nervous system, and at the individual muscle fibre, can be directly altered by elevations in local muscle and core temperature.
Fatigue is a multi-modal phenomenon, so the purpose of this chapter is to survey the direct effects
of heating on neuromuscular function, from the electrophysiology of isolated motor units undergo-
ing stimulation, to isometric and dynamic contractions of isolated muscles, and through to whole-
body exercise. A second objective is to briefly summarize the major methods employed by thermal
and muscle physiologists to investigate muscle function, and highlight some of the limitations and
challenges with current knowledge and technology.

Hyperthermia increases the physiological strain on the body, and exercise in hot
and even temperate environments can severely impair exercise capacity, with significant decreases in tolerance time to exhaustion. At the same time, cooling
strategies prior to or during exercise have become common in both athletic (e.g.
pre-cooling in rowing) and occupational (e.g. microclimate cooling systems
underneath protective garments) applications [1]. Traditionally, exercise in the
heat has been assumed to be primarily limited by cardiovascular constraints, with
reduced blood volume along with reduced venous return from increased dilation
of peripheral vascular beds for thermoregulation leading to cardiovascular collapse. Additionally, thermal strain is usually accompanied by changes in blood
distribution priorities throughout the body, and an impairment of blood pressure

or critical levels of blood flow to different organs such as the brain and the splanchnic tissues may accelerate fatigue and precipitate exhaustion.

The numerous observations of a consistent terminal core temperature (T_c) at the point of voluntary exhaustion, despite a range of physiological manipulations, is strongly indicative that core temperature has a direct effect on exercise capacity. Studies on the underlying mechanisms have generally focused on isolating specific systems and determining whether significant impairment was observed at either specific core temperatures and/or the point of exhaustion. The potential mechanisms for this relationship include cognitive arousal, cerebral blood flow, neurotransmitter concentrations and activity, metabolic alterations, and gastric permeability leading to endotoxaemia [2]. Ultimately, however, the cumulative effect from alterations in these other physiological systems is the failure of the neuromuscular system to sustain contractile force, precipitating the cessation of exercise. This form of fatigue induced by hyperthermia has been observed from a variety of perspectives, ranging from individual motor units through to whole-body exercise. Fatigue is a multi-modal phenomenon (for a review of muscle fatigue, see [3]), and it should be evident that hyperthermia may affect muscle function through different physiological mechanisms depending on the task being performed or tested. The difficulty, from a mechanistic perspective, is understanding the nature and site(s) of impairment in neuromuscular function due to heat. One challenge is that force production is a dynamic function between both central drive, its propagation through the neural pathways, and the characteristics of the muscle itself. Each can be altered by direct temperature effects, and the feedback from thermal afferents may also influence neuromuscular activation.

The purpose of this chapter is to survey the potential effects of hyperthermia along the continuum of neuromuscular function, from the individual muscle fibre through to whole-body force output and activation. Specifically, what are the direct effects of temperature on neuromuscular function, and how might this contribute to lower exercise capacity or tolerance during whole-body exercise in hot environments? Additionally, we will explore the different neuromuscular contributions to the premature cessation of whole-body exercise, evidenced by an inability to sustain a set work intensity or a voluntary lowering of self-selected pace. In addition, numerous techniques have been employed to quantify muscle function during exercise in the heat, and this review will attempt to provide a basic overview of these techniques, along with specific methodological issues and future applications.

Temperature and Muscle Electrophysiology

Temperature variation can affect the transmission of neural signals along the peripheral nervous system and down to the individual motor unit and muscle fibre.

Each individual component within this pathway may have different mechanisms or sensitivities to heat, resulting in a complex interaction. To date, the majority of clinical studies have focused on the effects of cooling on isolated muscle, and generally at levels much lower than typical of the large muscles during exercise [4].

Temperature and Action Potentials

In one of the few studies to systematically test the effects of heat on the electrophysiology of human muscles in vivo, the first dorsal interosseous (FDI) muscle of the hand was tested at stabilized skin temperatures of 32 and 42°C [5]. This was achieved via water immersion of the hand, and a close linkage between skin and muscle temperature of this thin and flat muscle was verified on one subject. The FDI proves to be a good model for testing temperature effects on muscle, due to its flat and thin shape, proximity to the skin surface, and also relative isolation from other large muscles. The study then attempted to systematically stimulate and analyze some of the major components of the neuromuscular system, including nerve conduction and activity across the neuromuscular junction. Activity and response of single motor units were also isolated to remove confounding factors from activity of multiple motor units of different morphologies. A summary of the major changes are presented in table 1.

One of the primary changes evident with temperature within the neuromuscular system is an alteration in conduction velocity of the nerves, and this can occur with both motor and sensory nerves. Impaired neural transmission is the primary cause of neuromuscular diseases associated with demyelination of the central (e.g. multiple sclerosis) or peripheral nervous system, and changes in conduction velocity can potentially alter neural transmission during exercise in healthy populations. Conduction velocity at 42°C increased for both motor and sensory nerves. Contributing to the changes in conduction velocity, the amplitude and duration of the action potential for both motor and sensory nerves decreased dramatically with heating, and is proposed to be due to alterations in sodium channel and possibly the potassium channel function with temperature. In this model, the activation and deactivation of these ion channels occur more rapidly with heating, with lower overall time for the channels to be open to permit charge influx and hence action potential amplitude. At extreme levels of temperature, conduction block can potentially occur, due to the channels opening and closing so rapidly that signal propagation ceases across nodes of the neuron. Decreasing core temperature by water immersion or cooling garments significantly improves the symptoms, muscle strength, and functional mobility of individuals with multiple sclerosis. However, as it relates to healthy individuals during exercise, local or systemic hyperthermia within physiological ranges should facilitate the propagation of neural signals from the brain along the peripheral nervous system to the muscles, and does not appear to limit exercise in the heat.

Table 1. Summary of patient data from electrophysiologic parameters at stabilized local skin temperatures of 32 and 42°C (average of all subjects) with stimulation of the first dorsal interosseous muscle

	32°C	42°C	Percent change
Motor amplitude, mV	11.4	8.4	−27
Motor duration, ms	2.6	2.1	−19
Forearm velocity, m/s	61	67.6	11
Sensory amplitude, µV	36.1	17.9	−50
Sensory duration, ms	0.72	0.52	−28
Sensory velocity, m/s	57.2	59.8	5
Decrement with 3-Hz repetitive stimulation			
Pre-exercise, %	4.2	1.7	
3 min post-exercise, %	2.0	2.3	
Single motor unit amplitude, µV	30	22	−26

Data from Rutkove et al. [4].

Upon arrival at the muscles, a neural signal must transfer from an electrical signal to a chemical signal across the neuromuscular junction. Temperature may affect the rate of this transmission through altering the production or activity of synaptic transmitters like acetylcholine, or through effects on its release and movement through the neural membrane. However, stimulation of the ulnar nerve and measurements at the first dorsal interosseous muscle did not produce significant declines in compound motor action potentials (CMAP) amplitudes with repetitive nerve stimulations at either 32 or 42°C skin temperature [5]. In addition, no changes were observed before and after 1 min of continuous exercise in either temperature condition. Overall, this suggests that the neuromuscular junction remains capable of full function regardless of temperature or sustained exercise, and likely does not contribute to exercise fatigue in the heat.

At the level of an individual motor unit, studied by isolating the FDI motor unit with the lowest threshold for stimulation, heating to 42°C skin temperature resulted in a significant decrease in amplitude of the action potential [5]. This decrease occurred progressively with time of heating, stabilizing after approximately 15–20 min. In turn, the amplitude steadily increased upon removal from heating and returned to baseline levels as the hand cooled. This gradual decrease and return of amplitude strongly suggests a direct effect of muscle temperature on the electrophysiological signal from an individual motor unit, and the magnitude of change was similar to that observed in CMAPs with stimulation of the whole FDI. In addition, the threshold voltage required to achieve stimulation did not appear to change with temperature.

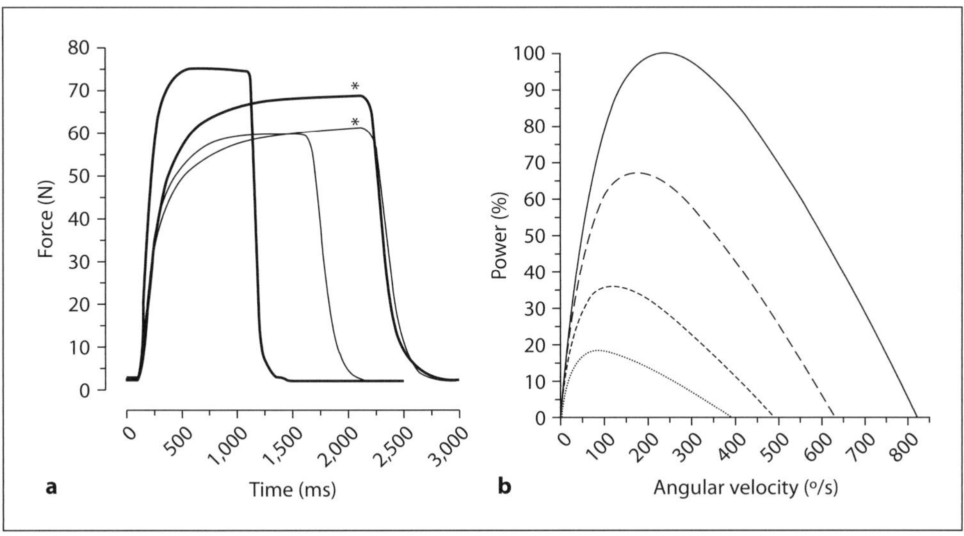

Fig. 1. **a** Examples of force records of fresh (thick traces) and fatigued (thin traces) adductor pollicis muscle, stimulated with 50 Hz at estimated muscle temperatures of 22 (∗) and 37°C. All traces were taken from the same subject. Reprinted from de Ruiter et al. [6]. Copyright Blackwell Publishing, used with permission. **b** Power/velocity curves of adductor pollicis muscle at calculated muscle temperatures of 37.1°C (full lines), 31.4°C (long dashes), and 25.6°C (short dashes), and 22.2°C (dotted line). All values were normalized with the maximal power production at 37.1°C equalling 100%. Reprinted from de Ruiter et al. [7]. Copyright Springer-Verlag, with permission.

In summary, in humans exercising in the heat, it appears that elevated local or whole-body temperatures does not appear to significantly impair neuromuscular transmission along the peripheral nervous system to the muscle fibre. At the muscle fibre itself, the amplitude of the action potential appears to be decreased by elevated temperature, likely via temperature-dependent effects of the ion channels within the fibre. Of course, it should be kept in mind that such studies involve brief stimulations rather than prolonged exercise, where the minor changes observed here may play a greater role with sustained, repetitive contractions. The next question to ask, then, is whether this change in signal amplitude actually affects the contractile characteristics and force capabilities of the muscle.

Temperature and Muscle Stimulation

To investigate the change in signal amplitude on contractile characteristics, De Ruiter et al. [6] also performed local heating and cooling of the hand, across a range of conditions from muscle temperatures of 22–37°C (fig. 1). Muscle temperature was again closely correlated to skin temperature for the flat and thin adductor pollicis muscle. Twitch force and maximal isometric tetanic force was similar with muscle temperature at 25, 31 or 37°C, but was significantly lower at 22°C compared to 37°C. The stimulation frequency required to obtain both maximal force generation and rate of force development progressively increased with rising temperature, as did the rate of force generation and relaxation. This likely reflects a more rapid excitation-contraction coupling at higher temperatures, for example in the release and uptake of calcium, resulting in faster twitches and higher stimulation rates required to achieve fusion.

An interesting addition to this study [6] was the testing of muscle function during and following a fatigue protocol, tested by imposing a string of stimulations to 90% isometric torque. In most parameters, the effects of fatigue were more pronounced at the higher muscle temperature of 37°C than at other temperatures. During the fatigue protocol, this included a greater decline in absolute force and also the rate of force production, while a progressive slowing of the relaxation rate with increasing temperature was also observed. Following the fatiguing protocol, the relative decline in maximal isometric torque was greater in the 37 and 31°C muscle temperature conditions than with 25 or 22°C temperatures, and the effects of fatigue were overall less marked at cooler temperatures. A similar temperature dependency is observed in the force/velocity relationship. Decreasing muscle temperature in the adductor pollicis resulted in a progressive decrease in maximal stimulated force and the maximal shortening velocity, resulting in power outputs at 22°C only ~19% that at 37°C [7]. Again, following a fatiguing protocol, the relative impairment in the force and velocity, and consequently power, was much greater at higher muscle temperatures. If extrapolated to whole-body exercise, such observations can suggest a higher force and power capability with elevated temperatures, but also a potential local muscular basis for greater development or onset of fatigue during exercise in the heat. To continue along the continuum of muscular function, the next step is to explore temperature effects on voluntary contractions of muscle groups.

Temperature and Muscle Capacity

Even at the level of electrophysiology of a single nerve or muscle, the effects of temperature on neuromuscular function can be difficult to isolate to a single mechanism or process. Relating to whole body exercise and testing the human muscle in vivo, one major consideration is thermal exchange between the tested muscle, other nearby muscles, and the body as a whole. Therefore, to understand the effects of temperature on neuromuscular function, it is important to separate the relative contributions of local temperature changes within an individual muscle versus that from core temperature within the brain or other regions of the body. That is because, independent of central temperature, changes in the peripheral properties of a heated muscle may contribute to the reduction in force output through imposing a need for increased central drive in order to maintain force production. This thermal exchange also presents methodological difficulties in partitioning the direct effects of local muscle temperature versus elevations in core temperature. This section will attempt to present literature emphasizing both local and whole-body temperature manipulations. Investigations with local muscle temperature manipulations are further complicated by potential differential effects based on the type of contraction and also the range of recording or measuring techniques. Therefore, discussion will be further separated into sustained isometric contractions and then to dynamic movements.

Local Muscle Temperature

Maximal torque during brief and voluntary contractions of large groups of muscles appear to be relatively unaffected by local thermal manipulations. Passive local cooling or heating of the quadriceps or biceps brachii did not affect peak torque with either isometric knee extensions [8] or elbow flexion [9], respectively. This may be a consequence of inadequate cooling or heating of a large muscle mass, however. With a more localized and isolated muscle such as the adductor pollicis, peak voluntary isometric torque was lower at 22°C than at 37°C [6]. However, the dynamics of temperature effects alter rapidly when engaging in prolonged contractions at either maximal or submaximal voluntary efforts. The same local muscle heating and cooling of the quadriceps resulted in a strong inverse relationship between time to fatigue and local temperature with sustained isometric knee extensions [8]. Such disparity between a relatively brief MVC versus sustained and submaximal contractions suggests that the ability to maximally recruit muscular force is independent of local muscle temperature.

Sustained contractions of large and multiple muscle groups, however, are influenced to a greater extent by thermal effects within the muscle. Substrate utilization shifts towards anaerobic glycolysis within the fibre, in turn increasing the rate of acidification. Furthermore, changes in excitation-contraction coupling with temperature within the fibre may increase the frequency of neural stimulation required to maintain a force output, again increasing ATP requirements. This relationship between electrical stimulation and mechanical activity of the muscle with temperature changes can be studied by comparing electromyographic and mechanomyographic (MMG) responses, respectively, with the latter generated by pressure waves produced by lateral expansion of active muscle fibres and recorded by accelerometers. While no changes in the root mean square (rms-EMG) were observed across cooling, thermoneutral, and heating conditions during isometric contractions, the

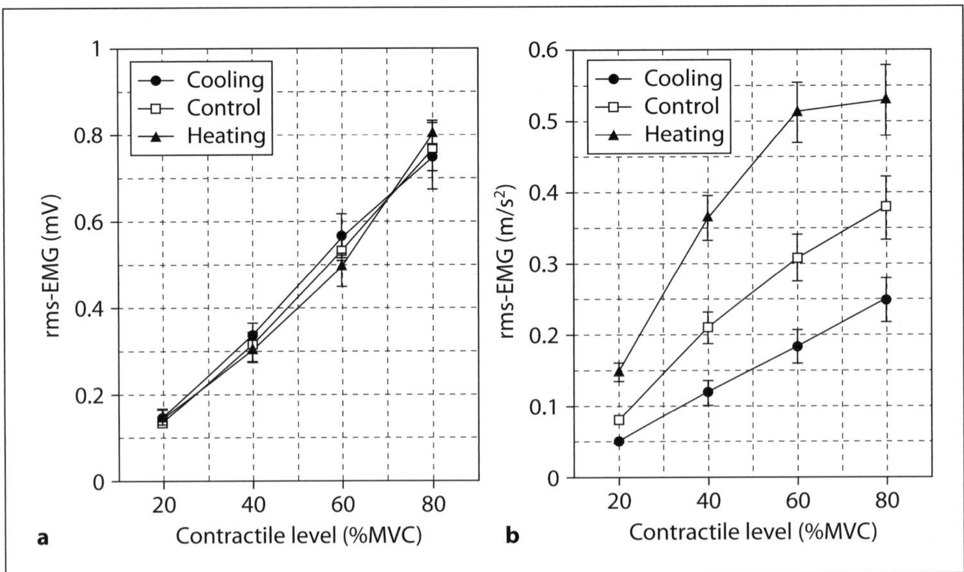

Fig. 2. Effects of temperature on root mean square amplitude of electromyogram (**a**, rms-EMG) and mechanomyogram (**b**, rms-MMG) during voluntary isometric contractions of different intensities. The values are presented as the mean with the standard error (SE) for ten subjects. The conditions of skin temperature are 28°C (cooling), 34°C (control), and 40°C (heating). Reprinted from Mito et al. [9]. Copyright Nauwelaerts Periodicals, with permission.

rms-MMG significantly increased with higher local temperatures over a range of submaximal forces (fig. 2), suggesting alterations in the mechanical contractile properties of the muscle with temperature [9]. The relationship between higher temperatures and a reduction in half-relaxation time and time to peak twitch force has been observed in maximal contractions on isolated muscles and with whole body hyperthermia, and the increased MMG may be reflective of a greater velocity or acceleration in the active muscle fibres in order to compensate.

Overall, one important caveat of many studies on local muscle thermal manipulations was that direct muscle temperature was typically not obtained, but was rather assumed or inferred from skin temperature. As outlined in the 'Methodological Issues' section, this assumption is dependent on skin temperature not having a significant effect on electrical signals, which may not be valid.

Core Temperature Effects

Studies eliciting hyperthermia via whole-body exercise in the heat have generally supported a dominant central impairment of neuromuscular activation above and beyond local temperature effects on muscle function. Following cycling-induced hyperthermia to 40.0°C, a greater decrease in maximal voluntary contraction force has been reported with both knee extension and hand grip compared to after cycling in a cool environment (core temperature 38.0°C) [10]. In the same study, interpolated twitch of the knee extensors also demonstrated a greater decline in voluntary activation at the point of exhaustion following whole-body hyperthermia. As cycling primarily utilized the lower body, the decrease in grip force of the non-exercised and 'passively' heated forearm was consistent with a decrease in central neuromuscular activation, though no activation measure of the forearm was obtained [10]. However, when the central activation ratio of the biceps brachii were actually measured following a similar cycling-induced hyperthermia protocol, no impairment in the non-exercised forearm flexors were reported with hyperthermia [11]. This differential activation pattern was proposed to be due to the central nervous system selectively regulating and reducing activation levels to specific skeletal muscles to protect against local tissue damage.

It is important to note that many of the key studies investigating various mechanisms underlying hyperthermia and exhaustion used exercise in hot environments to increase core temperature, and then compared the desired mechanism at baseline and at the point of voluntary exhaustion. While practical and of high ecological validity, two limitations may exist with this approach when exploring the direct effects of temperature per se. Firstly, the high cardiovascular strain and exercise-induced changes may influence the actual mechanism being studied and limit the isolation of temperature as a primary variable. Secondly, the lack of data as the individual becomes progressively hyperthermic provides no information on whether these changes occur progressively with increasing temperature or only upon attainment of a critical threshold temperature.

Some researchers have turned towards a passive heating model to further investigate temperature effects on neuromuscular activation. Morrison et al. [12] investigated the influence of hyperthermia on fatigue using a passive heating and cooling protocol, with isometric maximum voluntary knee extension and central activation measured from rest (~37.5°C) to 39.5°C at 0.5°C intervals, and then again at 0.5°C intervals as the individual was passively cooled back to 38.0°C. Using this model, hyperthermia was attained at a heart rate reserve of only 65%, and comparisons could be made at similar core temperatures with both warm and cool skin temperatures. Both maximal voluntary contraction torque and central activation of the knee extensors were progressively decreased with increasing rectal temperature. Secondly, it appears that core temperature was the primary thermal input causing hyperthermia-induced neuromuscular impairment, since when the skin was rapidly cooled (by ~8°C) and core temperature held stable at ~39.5°C, there was no recovery of MVC or voluntary activation. Furthermore, force and voluntary activation levels progressively returned to baseline values upon core cooling,

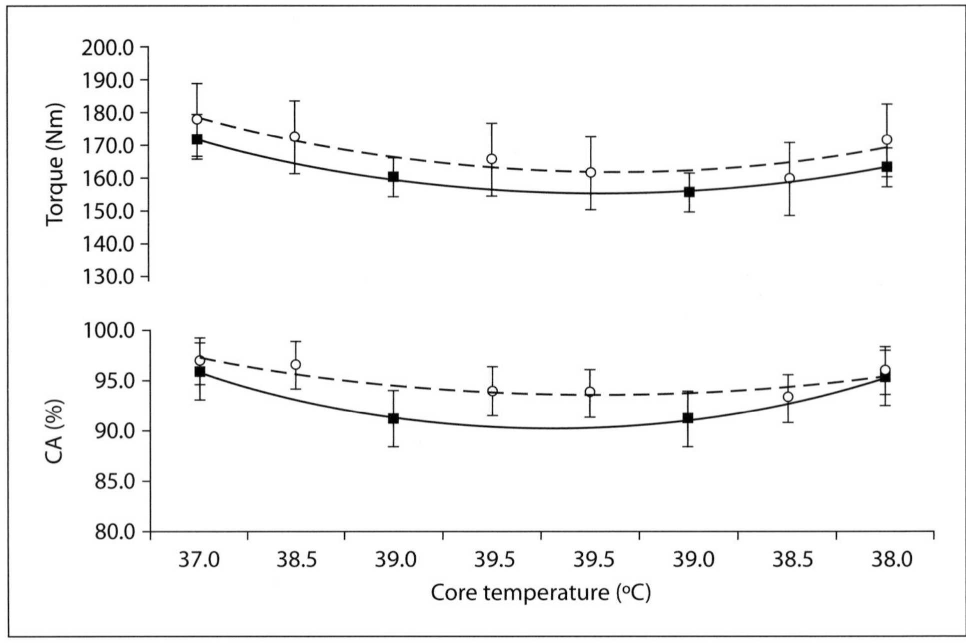

Fig. 3. Resting maximal voluntary contraction (MVC) torque and voluntary activation for heated (■) and thermoneutral (○) soleus muscle during passive heating from rectal temperatures of 37.2–39.5°C and then passive cooling to 37.9°C. Values are means (SD). Significant quadratic trends are present for heated (solid lines) and thermoneutral (dashed lines) muscle for both torque and voluntary activation. Reprinted from Thomas et al. [13]. Copyright American Physiological Society, with permission.

indicating that the ability to activate the muscle and produce force was not depressed as a result of fatigue accumulating over the protocol, but likely directly influenced by body core temperature. However, besides not measuring muscle temperature directly, another limitation was that the entire body, including the tested muscle, was theoretically heated and cooled following the same pattern as the whole body. Therefore, it was not possible to rule out the effects of local muscle temperature changes on the observed force and activation patterns.

To further separate local muscle and core temperature effects, Thomas et al. [13] performed the same passive heating-cooling protocol as Morrison et al. [12] while the soleus of one calf was kept cooled with an ice pack throughout the heating phase, with the other leg serving as a control that followed the heating-cooling pattern of the whole body (fig. 3). Deep muscle temperature of the soleus was successfully maintained at baseline levels of ~35°C in one leg, while the control leg reached a muscle temperature of 38.7°C at a core temperature of 39.0°C. Similar patterns of progressive impairment with increasing core temperature and progressive return to baseline with core cooling were observed in both the thermoneutral and the contralateral (heated and cooled) soleus, further supporting a primarily central impairment of neuromuscular activation rather than local muscle temperature [13].

Consistent with these passive heating studies, isometric force production and voluntary activation of the elbow flexors has been shown to decrease when core temperature was passively elevated to 38.5°C compared to 37.0°C [14]. Unique to this study was the use of transcranial magnetic stimulation, which demonstrated no change in motor cortical excitability with either thermoneutral or warm core temperatures. Coupled with an approximate 20% increase in peak relaxation rate of the elbow flexors during hyperthermia, the authors concluded that descending voluntary drive, at a level below the motor cortex, was not able to compensate for changed local muscle properties with hyperthermia. This, however, seems to conflict with the proposal that core temperature, presumably through action at the brain, is a primary driver for neuromuscular activation changes or voluntary fatigue.

In summary, it appears that neuromuscular impairment, at least in isolated isometric contractions at maximal activation, is focused at a point above changes in the properties of the muscle due to local temperature effects. Therefore, when determining exercise fatigue in the heat, central mechanisms appear to dominate, and emphasis for whole-body exercise needs to be on maintaining cooling of the entire body rather than individual muscles. However, the impairment due to core temperature elevation may not be at the level of the motor cortex itself, but at an intermediary level. The extent to which studies utilizing isometric and maximal contractions may transfer to voluntary fatigue with dynamic and whole-body exercise at submaximal force levels is arguable, however. The effects of temperature on dynamic movements will therefore be explored in the following section.

Dynamic Movements

Performance in most sporting situations is not reliant on maximal or sustained isometric contractions. Rather, the reliance on power, or rapid and dynamic force production places emphasis on parameters such as the rate of muscle force production and relaxation, along with the force-velocity relationship. Changes in action potential amplitude with stimulation (see 'Electrophysiology' section) suggest that the contractile characteristics of the muscle can also alter with temperature. While maximal isometric contractions provide insight into the limits of muscle activation, translation to whole-body exercise and voluntary fatigue may be tenuous due to the different nature and intensities of contractions. Compounding the interpretation of results are methodological issues such as differences in types of contractions and quantification of muscle activity.

Building on isometric tests at exercise-induced hyperthermia, Ftaiti et al. [15] performed maximal voluntary contractions before and following treadmill running in the heat to a point near volitional exhaustion (heart rate 196 bpm, tympanic temperature 40°C). The test battery involved both maximal isometric contractions and also isokinetic contractions ranging from 60 to 240°·s⁻¹. In addition, a 20 s endurance test of MVCs at 240°·s⁻¹ was used to explore the effects on prolonged dynamic exercise. The isometric data supported the decreased maximal torque with hyperthermia. Interestingly, the dynamic contractions followed a gradient depending on speed of contraction. While decreased torque and EMG activity were found with slower (60°·s⁻¹) isokinetic contractions, no impairment was observed at fast (240°·s⁻¹) contraction speeds during either the maximal or endurance tests. Further interpretation of the data is confounded by the deliberate design of 2% body mass loss over the exercise, such that hydration changes within the muscle may have impacted muscle responses.

In contrast, voluntary activation following exercise-induced hyperthermia by cycling was initially at pre-hyperthermia levels during maximal isokinetic contractions of the knee extensors, but decreased by 5.8–8.5% at the end of 25 repeated contractions [16]. Such decreases, however, were not observed in elbow flexion, suggesting that the decrement may be primarily due to exercise-related changes in the muscles rather than hyperthermia per se. Unfortunately, no muscle temperatures were measured in either study, again making direct comparisons of temperature effects difficult. Nevertheless, such studies highlight the potential role of different mechanisms and the difficulty in extrapolating data from isometric to dynamic contractions and whole-body exercise.

Temperature and Whole-Body Exercise

Thus far, potential sites for local muscle impairment have been presented, and it should be evident that temperature can significantly impair the contractile characteristics of muscles. In addition, hyperthermia also influences the body's ability to centrally and maximally recruit muscle. This section will briefly explore the research on whole-body exercise capacity in the heat, primarily with a view of elucidating temperature effects on neuromuscular function. However, because these effects seem to be contingent on the exercise modality, the reader is referred to the chapter by Tucker [this vol., pp. 26–38] for a more detailed discussion of this aspect.

Individual components of muscular performance may be altered to different extents by thermal stress. Following 30 min of sauna exposure that raised oral temperature by 2.5°C, trained weightlifters experienced significant reductions in muscular endurance with both bench press and leg press exercise [17]. However, muscular strength data is equivocal, with an impaired 1-RM in leg press but no difference with bench press. Finally, muscular power increased, with a significant improvement in vertical jump height. Broadly, these observations are consistent with the unchanged maximal voluntary torque but decreased tolerance times with sustained isometric contractions after surface heating of the quadriceps [8]. Also the increased muscular power is supported by the higher force and velocity of stimulated contractions with heating of the adductor pollicis [6]. One trade-off for higher power outputs with heating appears to be a faster onset of fatigue. With Wingate sprints on a cycle ergometer, the fatigue index was significantly higher in a 30°C than 18.7°C environment [18]. Whether this decrease is due primarily to peripheral or central factors is unclear. However, the accelerated fatigue in the heat supports the greater rate of force decrease with repeated stimulations of the warmed muscle [6].

Further support for a higher muscular power in the heat was provided by higher peak and mean power outputs during Wingate cycling sprint tests after 30 min exposure in warm (30°C) environments compared to normothermic (18.7°C) temperatures [18]. As resistance is constant with a Wingate test, the higher power outputs were concluded to be due to the higher pedalling cadences in the heat. Unfortunately, no thermal measurements or neuromuscular data was investigated in this study, such that firm conclusions concerning their relationship in this study remain tenuous. However, the higher cadences and power output in the heat may be due to higher temperature selectively optimizing the force/ velocity and power profiles of type I muscle fibres. In subjects with a range of type I percentage in the vastus lateralis ranging from 41 to 85%, muscle temperature was tested at baseline or manipulated to 26 and 39°C via immersion of the lower limb, with 5-second maximal sprints then occurring on an isokinetic cycle ergometer at 60, 110 and 140 rpm. A significant fibre-type dependency was observed with heating, such that individuals with higher type I percentages experienced greatest improvements in power output [19]. Thus, at each pedalling rate, the Q_{10} for power increased linearly (e.g. Q_{10} at 110 rpm was 1.17 with 40% type I fibres, and extrapolated to 1.61 for 100% type I fibres). In contrast, individuals with high proportion of type II muscle fibres experienced much lower to no improvements in power outputs with heating.

Quantifying Neuromuscular Function

As we have seen in the previous sections, many different techniques are employed in investigating neuromuscular function, and one challenge in understanding the effects of temperature is the comparison and extrapolation across different studies, each utilizing different muscle actions and methodologies. This section will briefly survey some of the major techniques utilized in studying neuromuscular function, many of which have been discussed above.

Electromyography

As outlined, the electrical activity of a muscle can be recorded and analyzed to provide an index of muscle recruitment and activation during stimulation or voluntary exercise. Specifically, as electromyography (EMG) reflects changes in the number of active muscle fibres or excitation rates, shifts in the EMG amplitude and frequency can indicate changing force output, recruitment strategy, and muscle fatigue. For example, during a prolonged maximal isometric contraction, the EMG amplitude falls progressively, likely due to decreases in motor unit firing rates and fewer active units. However, during repetitive or sustained submaximal contractions there can be a rise in EMG amplitude as more muscle fibres are recruited. Changes in EMG spectral frequency during a contraction have also been used to identify fatigue. A shift to lower frequencies occurs and this is thought to be caused by a lower conduction velocity. While a useful broad measure of muscle activity, it can be difficult to differentiate between central and peripheral factors underlying these changes with fatigue when used with whole-body exercise.

In order to quantify maximal force output, one important consideration is the appropriate normalization of EMG signals. Due to variables such as electrode placement in repeated trials, the absolute EMG amplitude may vary, requiring a stable reference maximum to be determined with each trial. A second necessity is that such normalization procedures should be reflective of the test itself. In this respect, it has been proposed that maximal isometric voluntary knee extensions can be a valid reference for cycling exercise despite the different body movements [20]. Such knee extensions produced much higher integrated EMG (IEMG) signals than the multi-joint movement from fixed seated pedal contractions at 60 and 108° knee angles and also from a one-revolution maximum on a cycle ergometer, but the IEMG/force relationship remained stable between knee extensions and one-revolution maximum.

Twitch Stimulations

Direct stimulation of the motor nerve bypasses the central nervous system to isolate the peripheral nervous system and muscle, and can target individual motor units through to whole muscle. Twitch characteristics will be altered if peripheral fatigue is present, and stimulation can be performed with different frequencies, durations, and intensities to investigate specific muscle properties. In turn, parameters such as peak force, contraction and relaxation times, and rates of force generation and relaxation can provide insight into the excitation-contraction coupling and the contractile characteristics of the muscle itself. For example, twitch force generation and relaxation rates provide information on the sarcoplasmic reticulum's capacity for Ca^{2+} release and uptake, respectively. Twitch characteristics of muscle can also be used to study the relative contributions of different muscle fibre types to contraction, as type I fibres have slower contraction and relaxation times and lower peak forces than type II fibres.

Compound Motor Action Potentials

At the level of the whole muscle, the compound motor action potential (CMAP, also termed the M-wave) is a summation of the action potentials elicited by a single maximal stimulus to the motor nerve of a muscle. The M-wave is recorded by EMG electrodes during twitch stimulation, and changes in the general properties of the M-wave can be used to indicate fatigue. For example, increases in M-wave duration with a decrease in peak-to-peak amplitude and total area are symptomatic of fatigue. As outlined above, more detailed analysis of the waveforms can be performed to investigate the underlying mechanisms of fatigue, such as changes in the neuromuscular junction or membrane excitability resulting from deficiencies in transport of Na^+ and K^+ across the sarcolemma and t-tubules.

Interpolated Twitch Technique

Voluntary activation refers to the central 'drive' to the muscular system, which in turn is propagated along the peripheral motor neurons to activate a contraction. Therefore, the level of voluntary activation can be an indication of whether a reduction in generated force is due to fatigue within the local muscle or a central inability to maximally activate the total motor neuron pool reaching that muscle. One of the dominant methodologies for assessing voluntary activation is the interpolated twitch technique, first developed by Merton [21] in 1954. The technique involves the subject performing a maximal isometric voluntary contraction, and then having a supramaximal stimulated contraction superimposed on top of the voluntary contraction. Thus, the interpolated twitch is an index of the proportion of motor units that are not fully recruited during a voluntary effort, and can

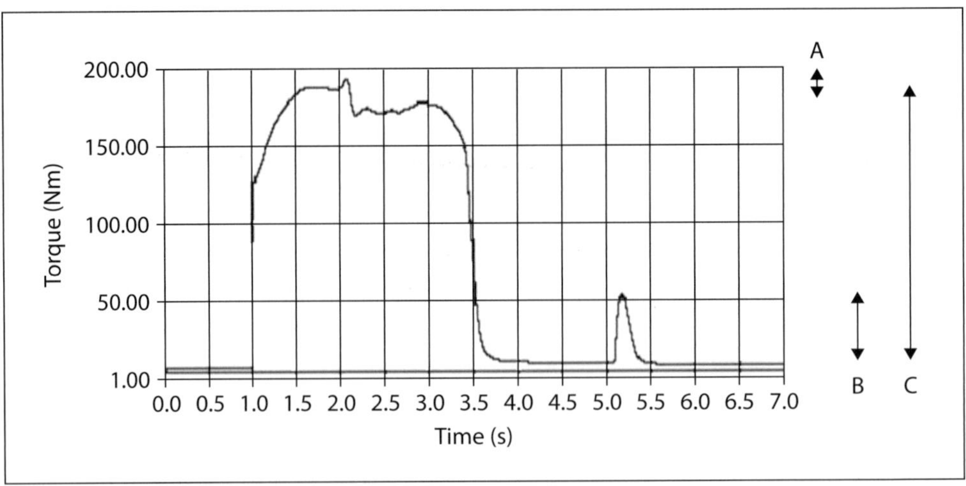

Fig. 4. Stylized representation of data from an interpolated twitch experiment, involving an isometric maximal voluntary contraction (MVC, C arrow) with a superimposed or interpolated twitch (A arrow) during the stable portion of the MVC. The same stimulus is then applied following brief relaxation of the muscle to obtain a potentiated twitch (B arrow). See text for explanation of using these values to calculate voluntary activation and central activation ratio.

provide information on the relative contribution of peripheral muscle function versus central activation. Two methods of calculating the relative effects from interpolated twitch (fig. 4) are: (a) voluntary activation (VA), and (b) central activation ratio (CAR).

Voluntary activation is calculated using the formula:

$$\text{VA} = (1 - \text{interpolated twitch amplitude/potentiated twitch amplitude}) \times 100\%$$

Central activation ratio is calculated using the formula:

$$\text{CAR} = \text{MVC amplitude/(IT amplitude} + \text{MVC)} \times 100\%$$

Though the existing work in thermophysiology has employed only maximal contractions, interpolated twitches can also be performed at submaximal contraction intensities. Some care needs to be taken in interpreting and extrapolating data from the interpolated twitch technique. A maximal voluntary contraction, especially of a large limb or joint, can recruit the primary agonist muscle but also potentially a large number of synergist muscles. In contrast, the evoked force from direct nerve stimulation may activate only the primary agonist muscle. Therefore, the greater musculature recruitment during voluntary contractions may not be an appropriate comparison with the evoked force and artificially lower the VA or CAR values. Another inherent limitation of the interpolated twitch technique is that it is restricted to isometric contractions, making results from such studies potentially difficult to extrapolate to dynamic or whole-body exercise.

Transcranial Magnetic Stimulation

The previous methodologies have primarily assessed peripheral neuromuscular function, with only indirect information on central neuromuscular activation. Recently, transcranial magnetic stimulation (TMS) has been employed in thermophysiology to assess the extent of excitability or reserve within the motor cortex itself [14]. Similar to the interpolated twitch technique, TMS superimposes a neural stimulation on top of a submaximal or maximal voluntary contraction. However, rather than stimulating the peripheral nervous system at the level of the motor nerve, TMS activates the motor cortex to more directly assess central drive. In addition to methodological difficulties in stimulating the correct area of the motor cortex, similar caveats of comparing voluntary with evoked force exist as with the interpolated twitch technique.

Methodological Issues

Effects of Temperature Manipulations on Muscle Temperature
One important issue when investigating the effects of thermal changes on muscle function is the actual effectiveness of a thermal manipulation on local muscle temperature, along with the potential presence of temperature gradients throughout a muscle. Many studies investigating thermal effects on neuromuscular function assume that actual nerve or muscle temperature can be correlated to skin temperature or core temperature, and also that the muscle as a whole is homogenous in temperature. While muscle and skin temperature can indeed be closely correlated for some muscles, notably flat and thin muscles such as the first dorsal interosseous muscle of the hand, this may not be appropriate for larger muscles or deeper muscles, which are at a greater distance from any externally applied thermal manipulations. Furthermore, given the wide variability in size and shape of muscles and also body composition (e.g. subcutaneous fat), potentially large differences in the accessibility of muscles to thermal changes may occur across individuals.

This is notable for passive cooling or heating manoeuvres, where external temperature manipulations may not result in significant muscle temperature changes, or may have uneven temperature changes throughout the muscle. Therefore, it becomes difficult to isolate the actual effects of temperature changes on neuromuscular function. Another potential issue with using skin temperature as an index of deep tissue temperature is the relatively rapid change upon removal from a manipulation. This may be especially prevalent for local cooling manoeuvres, which may not be representative of deep muscle temperature. Intramuscular temperature recordings demonstrate a high gradient for temperature throughout the rectus femoris muscle when cooled for 20 min using surface ice-packs, with significant cooling occurring down to only 1 cm within the muscle but no changes at 2 and 3 cm depth [22]. This can likely be attributed to the relatively large amount of deep blood flow from the rest of the non-manipulated and warm body compared to the localized cooling. Indeed, upon removal of the ice pack, skin temperature rapidly rose back to levels near baseline, while temperatures at 2 and 3 cm depth decreased slightly, suggesting conductive heat exchange within the muscle. Overall, such observations highlight the difficulty in isolating or manipulating the temperature of individual muscle tissue in vivo, especially when muscle temperature is not directly recorded.

Effects of Muscle Activity on Temperature

Muscle activity itself will increase tissue temperature independent of any external or internal manipulations, and the degree of heat production will depend on muscle composition and contraction intensity. Maximal rate of temperature change is closely correlated to the fraction of type II fibres in the muscle, such that the biceps brachii, with a higher proportion of type II fibres, would be expected to have higher heat generation (55%, 1.46°C·min^{-1} at MVC) than a low-type II muscle such as the soleus (24%, 0.5°C·min^{-1}). Such temperature changes could become a significant factor with prolonged contractions, especially with repetitive fatiguing protocols. Adding to the difficulties in defining temperature effects is thermal exchange within the muscle itself and with the rest of the body. Isolated concentric knee extension exercise resulted in a 0.55°C increase in oesophageal temperature after 15 min, along with an increased vastus medialis temperature ranging from 2.00–3.20°C depending on distance from the femur [23]. Temperature gradients of 0.8–1.1°C were also evident within the muscle during both rest and recovery periods, with a sustained oesophageal temperature increase throughout recovery. In this way, it becomes difficult to extrapolate and isolate local muscle temperature effects with that of the body as a whole and vice versa. This is further exacerbated in dynamic movements involving many muscles, each of which may have wide variability in muscle temperature and also exchange heat with each other. Regardless, it appears critical that local temperature of the tested muscle(s) be directly measured whenever possible. While invasive intramuscular temperature may be difficult with whole-body exercise or field studies, some initial reports suggest that, at rest in thermoneutral environments, skin temperature insulated using neoprene disks of varying thicknesses may potentially correlate to intramuscular temperature [24].

Effects of Skin Temperature on Electromyography

The importance of muscle temperature standardization in clinical diagnostic use of EMG is fairly well-recognized. To permit comparison across tests and individuals, clinicians typically try to maintain an assumed muscle temperature of 34–36°C and 32–34°C in the upper and lower body, respectively. These form standardized conventions rather than 'optimal' conditions for EMG responses. In exercise science, electromyography is commonly employed as an index of muscle activity at levels of research ranging from isolated muscle through to whole-body exercise. In contrast to clinical situations, wide variability in muscle temperature is common with both exercise and heat exposure, making comparisons with 'standard' test results difficult. Also, as discussed above, skin temperature may not be reflective of actual temperature deep within the muscle.

Furthermore, the use of surface EMG on the skin assumes that skin temperature does not influence the quality or nature of the electrical signal, but this does not appear to be the case. Concentric contractions at ambient temperature ranging 14–30°C did not alter core temperature but skin temperature ranged from 21.7–32.9°C. When analyzing EMG of the soleus, cool skin temperatures resulted in a doubling of EMG amplitude (73–135 μV) with a concomitant reduction in mean EMG power frequency from 142 to 83 Hz [25]. A similar observation of reduced EMG to force relationship was observed with isometric knee extensions from 10–100% MVC over a 90-min exposure to 10, 23 and 40°C air temperatures [26]. Unfortunately, as with many studies on temperature effects and muscle function, direct muscle temperature was not measured in either study, such that it becomes difficult to determine whether such changes are solely due to an artefact from skin temperature differences or due to actual muscle temperature itself. However, given the lack of deep muscle temperature effects from ice pack application [22], it is unlikely that the mild ambient temperature range would significantly alter deep muscle temperature.

The changes in surface EMG signal due to skin temperature may be due to a number of methodological or physiological factors [26]. Electrode resistance did not vary across either temperature conditions or over time, such that any differences in conductivity or viscosity of the electrode or gel would not be responsible for the observed EMG changes. However, sweating, simulated by maintaining a wet electrode throughout 30 min exposure to 23°C, resulted in a higher EMG power than dry electrodes. In addition, subcutaneous fluid changes from temperature manipulations (e.g. increased shunting of blood to the peripheries with heating) may alter skin conductivity. Finally, changes in nerve conduction velocity with temperature may also lead to alterations in the EMG signal. Overall, it appears important to further investigate the bases of EMG signal alterations due to skin temperature, and to models such effects when employing EMG in exercise studies in the heat.

Limitations with Whole-Body Exercise

Testing systemically, using whole-body heat exposure and exercise, poses several potentially major methodological pitfalls that can confound data interpretation. First and foremost, whole-body exercise is inextricably intertwined with psychophysiological factors that alter an individual's drive to exercise. In turn, this is influenced by complex interactions involving extrinsic factors (e.g. motivational changes due to lab tests versus more realistic field trials, solo versus group settings) and also intrinsic factors stemming from physiological afferent feedback (e.g. thermal, muscular, and circulatory discomfort).

One expression of this methodological limitation is that, to minimize intra- and inter-individual variability and to ensure high levels of exertion and core temperature are achieved, the majority of studies on exercise-induced hyperthermia are largely restricted to highly fit and trained individuals, predominantly healthy young adult males. For example, only a few studies on tolerance to uncompensable heat stress have deliberately targeted subject recruitment towards those with low aerobic fitness, and even these subjects are in the middle to higher range of the population as a whole. Therefore, while of relevance to athletic applications, it may be difficult to transfer current research to the general population faced with heat waves or even the majority of workers in occupational settings [2].

With either whole-body passive exposure to heat stress or exercise in a hot environment, it becomes important to differentiate the effects of temperature from that due to hydration changes. Sweating and fluid compartment shifts can influence cardiovascular and hormonal responses beyond directly altering muscle contractile characteristics. Thirst and ingested volume can also be a significant behavioural feedback affecting discomfort and motivation. Methodologically, however, such separation can be very difficult to achieve with heating via active exercise or passive heat exposure. Even when a specific hydration strategy is implemented, the lag time from gastric emptying and absorption results in unavoidable delays, producing dynamic shifts in fluid compartments. A potential therefore exists for the use of intravenous rehydration during exercise to accelerate fluid replacement and also to control for perceptual effects due to different oral hydration strategies (e.g. volume, composition, temperature).

Conclusions

The effects of elevated muscle or body temperature can elicit fatigue through directly altering the capacity of the neuromuscular system to generate force. This can occur from the central activation and recruitment of force, the conduction of signal along the central and peripheral nervous system, through to contractile characteristics of individual motor units. Numerous techniques have been employed in this field of study, each targeting different levels of organization of the system and each having limits to the extrapolation of data. Overall, an integrated view suggests that hyperthermia can reduce the central drive or activation of muscle force, and also alters a muscle towards higher rates of force generation but potentially faster rates of force decline and fatigue.

References

1 Marino FE: Methods, advantages, and limitations of body cooling for exercise performance. Br J Sports Med 2002;36:89–94.
2 Cheung SS, Sleivert GG: Multiple triggers for hyperthermic fatigue and exhaustion. Exerc Sports Sci Rev 2004;32:100–106.
3 Enoka RM, Duchateau J: Muscle fatigue: what, why and how it influences muscle function. J Physiol (Lond) 2008;586:11–23.
4 Rutkove SB: Effects of temperature on neuromuscular electrophysiology. Muscle Nerve 2001;24: 867–882.
5 Rutkove SB, Kothari MJ, Shefner JM: Nerve, muscle, and neuromuscular junction electrophysiology at high temperature. Muscle Nerve 1997;20:431–436.
6 De Ruiter CJ, Jones DA, Sargeant AJ, De Haan A: Temperature effect on the rates of isometric force development and relaxation in the fresh and fatigued human adductor pollicis muscle. Exp Physiol 1999;84:1137–1150.
7 De Ruiter CJ, De Haan A: Temperature effect on the force/velocity relationship of the fresh and fatigued human adductor pollicis muscle. Pflügers Arch 2000;440:163–170.
8 Thornley LJ, Maxwell NS, Cheung SS: Local tissue temperature effects on peak torque and muscular endurance during isometric knee extension. Eur J Appl Physiol 2003;90:588–594.
9 Mito K, Kitahara S, Tamura T, Kaneko K, Sakamoto K, Shimizu Y: Effect of skin temperature on RMS amplitude of electromyogram and mechanomyogram during voluntary isometric contraction. Electromyogr Clin Neurophysiol 2007;47:153–160.
10 Nybo L, Nielsen B: Hyperthermia and central fatigue during prolonged exercise in humans. J Appl Physiol 2001;91:1055–1060.
11 Saboisky J, Marino FE, Kay D, Cannon J: Exercise heat stress does not reduce central activation to non-exercised human skeletal muscle. Exp Physiol 2003;88:783–790.
12 Morrison S, Sleivert GG, Cheung SS: Passive hyperthermia reduces voluntary activation and isometric force production. Eur J Appl Physiol 2004;91: 729–736.
13 Thomas MM, Cheung SS, Elder GC, Sleivert GG: Voluntary muscle activation is impaired by core temperature rather than local muscle temperature. J Appl Physiol 2006;100:1361–1369.
14 Todd G, Butler JE, Taylor JL, Gandevia SC: Hyperthermia: a failure of the motor cortex and the muscle. J Physiol (Lond) 2005;563:621–631.
15 Ftaiti F, Grelot L, Coudreuse JM, Nicol C: Combined effect of heat stress, dehydration and exercise on neuromuscular function in humans. Eur J Appl Physiol 2001;84:87–94.
16 Martin PG, Marino FE, Rattey J, Kay D, Cannon J: Reduced voluntary activation of human skeletal muscle during shortening and lengthening contractions in whole body hyperthermia. Exp Physiol 2005;90:225–236.
17 Hedley AM, Climstein M, Hansen R: The effects of acute heat exposure on muscular strength, muscular endurance, and muscular power in the euhydrated athlete. J Strength Cond Res 2002;16:353–358.

18 Ball D, Burrows C, Sargeant AJ: Human power output during repeated sprint cycle exercise: the influence of thermal stress. Eur J Appl Physiol 1999;79:360–366.

19 Sargeant A, Rademaker A: Human muscle power in the locomotory range of contraction velocities increases with temperature due to an increase in power generated by type I fibres. J Physiol (Lond) 1996;491:128P–129P.

20 Hunter AM, St Clair Gibson A, Lambert M, Noakes TD: Electromyographic (EMG) normalization method for cycle fatigue protocols. Med Sci Sports Exerc 2002;34:857–861.

21 Merton PA: Voluntary strength and fatigue. J Physiol (Lond) 1954;123:553–564.

22 Enwemeka CS, Allen C, Avila P, Bina J, Konrade J, Munns S: Soft tissue thermodynamics before, during, and after cold pack therapy. Med Sci Sports Exerc 2002;34:45–50.

23 Kenny GP, Reardon FD, Zaleski W, Reardon ML, Haman F, Ducharme MB: Muscle temperature transients before, during, and after exercise measured using an intramuscular multisensor probe. J Appl Physiol 2003;94:2350–2357.

24 Brajkovic D, Ducharme MB, Webb P, Reardon FD, Kenny GP: Insulation disks on the skin to estimate muscle temperature. Eur J Appl Physiol 2006;97:761–765.

25 Winkel J, Jorgensen K: Significance of skin temperature changes in surface electromyography. Eur J Appl Physiol Occup Physiol 1991;63:345–348.

26 Bell DG: The influence of air temperature on the EMG/force relationship of the quadriceps. Eur J Appl Physiol Occup Physiol 1993;67:256–260.

Stephen S. Cheung, PhD
Environmental Ergonomics Laboratory
Department of Physical Education and Kinesiology, Brock University
500 Glenridge Avenue, St. Catharines, ON L2S-3A1 (Canada)
Tel. +1 905 688 5550/5662, Fax +1 905 688 8364, E-Mail stephen.cheung@brocku.ca

Marino FE (ed): Thermoregulation and Human Performance. Physiological and Biological Aspects.
Med Sport Sci. Basel, Karger, 2008, vol 53, pp 61–73

Intestinal Barrier Dysfunction, Endotoxemia, and Gastrointestinal Symptoms: The 'Canary in the Coal Mine' during Exercise-Heat Stress?

G. Patrick Lambert

Department of Exercise Science, Creighton University, Omaha, Nebr., USA

Abstract

Reduced intestinal blood flow and high intestinal temperatures during exercise-heat stress can lead to intestinal barrier dysfunction. Such dysfunction may increase intestinal permeability to endotoxin. During exercise-heat stress, intestinal barrier dysfunction and endotoxemia can produce gastrointestinal symptoms and increased production of pro-inflammatory cytokines. Such problems may be a warning sign ('canary in the coal mine') for the onset of exertional heat stroke. Failure to heed such a warning may culminate in problems indicative of exertional heat stroke such as circulatory collapse and multiple organ failure. Prior exposure to exercise-heat stress may, however, be a protective mechanism.

If the combined environmental and metabolic thermal load exceeds the human body's ability to dissipate the load during exercise-heat stress, severe hyperthermia can occur. This can lead to potentially serious heat-related problems, including exertional heat stroke. The intent of this chapter is to bring attention to the mounting evidence that indicates that the gastrointestinal (GI) tract, and especially the intestinal barrier, may be one of the first areas negatively affected by exercise-heat stress. Furthermore, endotoxemia caused by dysfunction of the intestinal barrier is likely related to GI symptoms experienced by many individuals under such conditions, and these may serve as a warning sign (i.e. 'canary in the coal mine') for impending problems such as exertional heat stroke if left unheeded.

The Intestinal Barrier and Intestinal Permeability

As previously described [1], the intestinal barrier is composed of physical factors such enterocyte membranes and tight junctions between enterocytes, along with an immunological defense system. Its function is to guard against passage of potentially dangerous substances from the intestinal lumen to the internal environment.

If any of the above factors are compromised, intestinal permeability may increase. Intestinal permeability is defined as the nonmediated diffusion of molecules >150 Da in size. Increased intestinal permeability can expose the body's internal environment to higher-than-normal amounts of toxic substances from the intestinal lumen. Such substances include endotoxin (i.e. lipopolysaccharide; LPS), a highly immunogenic component of the outer cell wall of Gram-negative bacteria. Increased intestinal permeability has been observed in a number of pathophysiological conditions including, but certainly not limited to, inflammatory bowel disease, severe hyperthermia, hemorrhagic shock, and severe psychological stress.

Intestinal permeability can be assessed in humans by the urinary excretion of ingested 'probes' such as the nondigestible, nonmetabolizable substances lactulose (molecular weight, 342 Da), polyethylene glycol (molecular weight, 400–4,000 Da), or ^{51}Cr-EDTA (molecular weight, 343 Da). The amount of the ingested dose that is excreted in the urine serves as an index of permeability for that size molecule. Molecules of the sizes indicated above can only permeate the intestinal barrier via the paracellular route (i.e. by passing through disrupted tight junctions), or by direct entry to the internal environment due to damaged epithelium (i.e. enterocyte necrosis). A smaller molecule such as rhamnose (molecular weight, 164 Da), transported both transcellularly and paracellularly, is also normally co-ingested with one of the larger molecules, such as lactulose. This is done to control for nonmucosal factors (e.g. GI transit, renal clearance) that may affect urinary excretion among experimental conditions. Thus, intestinal permeability frequently is assessed by the lactulose-to-rhamnose (L/R) urinary excretion ratio.

LPS itself may also serve as an indicator of intestinal barrier dysfunction, since gut-derived LPS is highly restricted from the internal environment due to both its large size and pathogenic nature. Thus, in the absence of other routes of entry to the circulation, LPS detected in the portal or systemic circulation indicates dysfunction of both the physical and immune barriers of the intestine.

Intestinal Barrier Dysfunction and Endotoxemia during Exercise-Heat Stress

In humans, Brock-Utne et al. [2] provided the earliest evidence that the intestinal barrier was being compromised during prolonged, strenuous exercise. In that study, 81% of the 89 randomly selected runners requiring admission to the medical

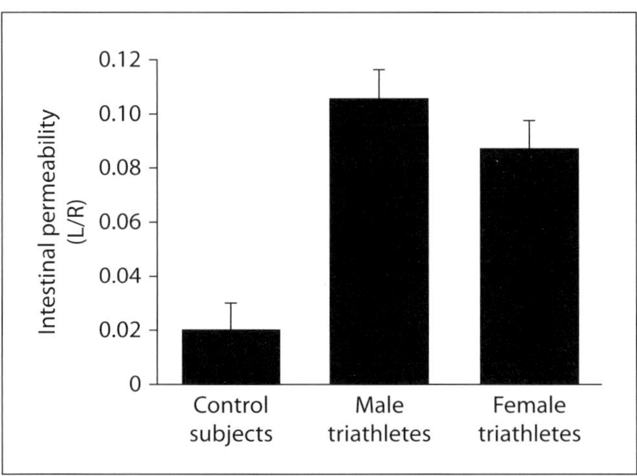

Fig. 1. Small intestinal permeability following the 1998 Hawaii Ironman Triathlon (3.84 km swim, 179 km bike, 42 km run). Values for male and female competitors were each significantly (p < 0.05) greater than for resting controls. Data from Lambert et al. [7].

tent following completion of the Comrades Marathon (89.4 km) were found to be clinically endotoxemic with 2% above the lethal level. At relatively the same time as that study was conducted, investigators from the same lab showed that heat stress in primates led to increased portal and systemic LPS concentrations [3]. This group subsequently found that administration of gut-cleansing antibiotics prevented endotoxemia from occurring under such conditions, providing evidence that endotoxemia occurring with heat stress is gut-derived [4].

A number of other investigators have subsequently reported increased intestinal permeability and/or endotoxemia during strenuous exercise and/or heat stress in humans. In 1991, Bouchama et al. [5] reported that LPS was elevated in all patients (n = 17) admitted to the hospital for heatstroke. In that study, rectal temperatures averaged 42.1°C upon admittance. In 1997, Pals et al. [6] found that running at 80% $VO_{2\,max}$ produced a significant increase in intestinal permeability (i.e. L/R) compared to running at 40 or 60% $VO_{2\,max}$. In that study, final rectal temperatures at 80% $VO_{2\,max}$ approached 40°C (mean = 39.6°C). In 1999 and 2000 respectively, Lambert et al. [7] and Jeukendrup et al. [8] reported intestinal barrier dysfunction following similar warm-weather, long-distance triathlons (3.8 km swim, 179 km bike, 42 km run). Figure 1 represents the results observed by Lambert et al. [7], in which both men and women subjects experienced significant increases in small intestinal permeability (i.e. L/R). The subjects studied by Jeukendrup et al. [8] were found to have significantly elevated plasma LPS concentrations.

Other physiological models have further supported changes in intestinal permeability with heat stress. In vitro, tight junction dysfunction was observed by Moseley et al. [9] using cultured MDCK cells (i.e. canine kidney tubule cells). In that study, heating induced increased paracellular permeability to larger molecules (i.e. mannitol; molecular weight, 180 Da) at 41.3°C. In vivo, Hall et al. [10] subsequently reported significantly increased LPS in the portal blood of heat-stressed rats with rectal temperatures of 41.5°C. In both in vivo and ex vivo (i.e. everted intestinal sac) models, Lambert et al. [11] reported that tissue temperatures of 41.5–42°C for only 60 min increases permeability to large molecules (i.e. FITC-dextrans; molecular weight, 4,000 Da). Dokladny et al. [12] also recently reported increased tight junction permeability at more modest, but longer periods of heating (41°C over 24 h) in cultured Caco-2 cells (i.e. colonic cell line).

Causes of Intestinal Barrier Dysfunction and Endotoxemia during Exercise-Heat Stress

Reduced Intestinal Blood Flow

At exercise intensities of 50–60% $VO_{2\,max}$ in a hot environment it has been shown that splanchnic blood flow (SBF) can decrease 50–60% in humans [13]. As indicated in figure 2, many of the mechanisms associated with intestinal barrier dysfunction during exercise-heat stress can stem from reduced intestinal blood flow. These include tissue hypoxia, ATP depletion, acidosis, and oxidative/nitrosative stress.

Tissue Hypoxia

Insufficient intestinal perfusion during heat stress can promote intestinal mucosal hypoxia. This was demonstrated in rats by Hall et al. [14] using the hypoxic marker [^{3}H]mizonidazole. Heat stress sufficient to raise rectal temperatures to 41.8°C resulted in 29% retention of the probe in the intestine and 80% retention in the liver. It has been further demonstrated in vitro (i.e. Caco-2 cells) that hypoxia leading to ATP depletion causes tight junction hyperpermeability [15]. Hall et al. [14] also found that hypoxia during heat stress likely causes tissue acidosis, and it is known that acidosis, under conditions of reduced blood flow, promotes increased intestinal permeability [16].

Oxidative and Nitrosative Stress

It is well documented that intestinal ischemia and ischemia-reperfusion increase the production of reactive oxygen species (ROS) which leads to epithelial damage [17]. This was demonstrated by Hall et al. [10] who observed that inhibition of ROS production during heat stress in rats reduces intestinal permeability to LPS. Hall et al. [18] further demonstrated that nitric oxide (NO) production increases

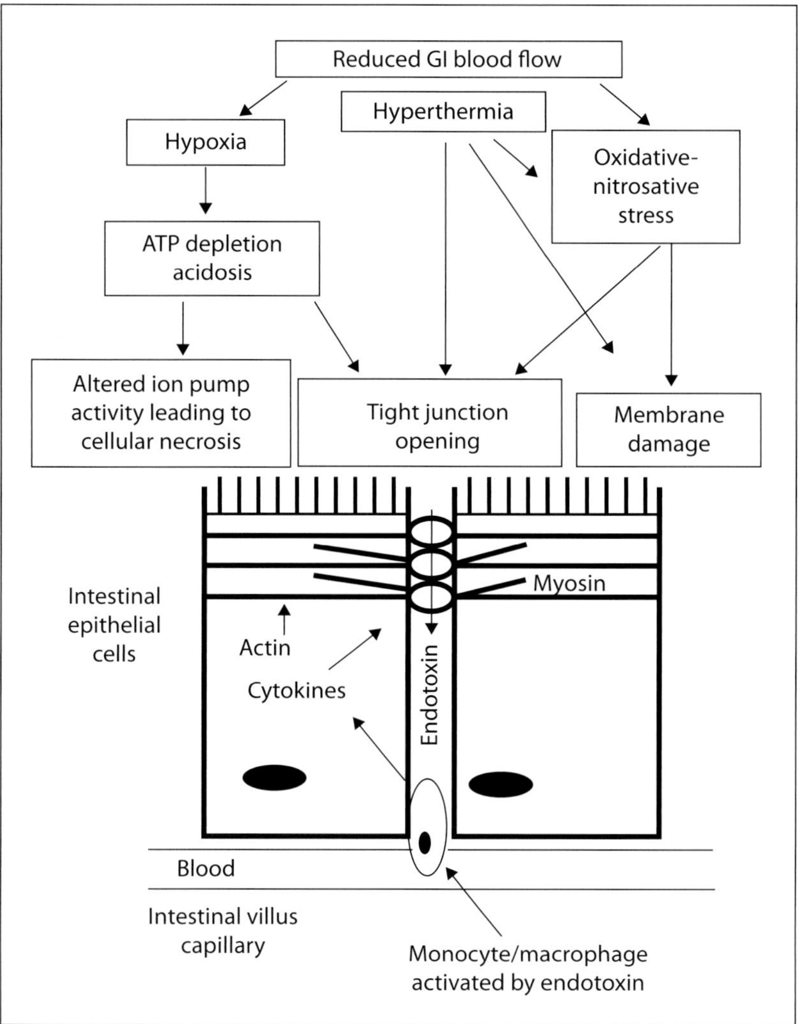

Fig. 2. Likely mechanisms promoting intestinal barrier dysfunction during exercise-heat stress. Reprinted from Lambert [1], with permission. Copyright 2004 American College of Sports Medicine.

during hyperthermia and it appears that high concentrations of NO promote increased intestinal permeability [19] and may be related to splanchnic vasodilation and circulatory collapse that occurs with severe hyperthermia [20].

Hyperthermia
It has been found in humans exercising to exhaustion in the heat that internal splanchnic temperatures (i.e. hepatic venous temperature) can reach 41.7°C. This is

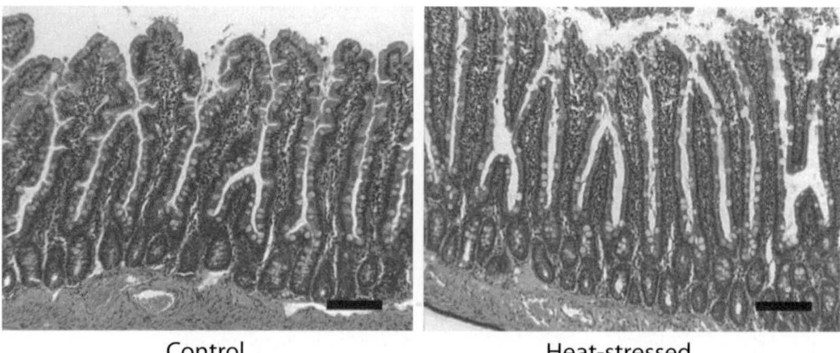

Fig. 3. Light micrographs of small intestinal villi from anesthetized control (37°C) and heat-stressed (41.5–42.5°C) rats. Heat stress results in epithelial damage (necrosis and sloughing of enterocytes) from villous tips. Bars = 100 μm. Reprinted from Lambert et al. [11], with permission. Copyright 2002 American Physiological Society.

1.5°C higher than rectal temperatures (i.e. 40.2°C) in the same subjects [21]. Based on such information, it is possible that individuals with rectal temperatures of 39–40°C have internal (e.g. intestinal) temperatures sufficient to produce tight junction permeability [9], membrane damage, and enterocyte death [11]. Such effects are shown in figures 3 and 4 in which core temperatures of 41.5–42.5°C for only 60 min caused severe intestinal epithelial and enterocyte membrane disruption.

Dehydration

Dehydration appears to exacerbate intestinal barrier dysfunction likely by further reducing intestinal blood flow. Accordingly, mesenteric flow has been found to be reduced by 50% in dehydrated rats forced to swim in warm water [22]. Furthermore, Lambert et al. [23] recently observed that fluid restriction during prolonged running results in significant increases in intestinal permeability.

Local Inflammatory Mediators

Passage of LPS into the internal environment following loss of tight junction integrity and/or enterocyte damage can lead to a local immune response via activation of T lymphocytes, monocytes, and tissue macrophages. This response involves increased production of pro-inflammatory cytokines such as tumor necrosis factor-α and interleukin-1, along with other immune modulators, such as interferon-γ, and NO. All of these substances are well-known to affect the intestinal mucosal barrier and can lead to even greater increases in permeability [24]. Thus, initial LPS permeation due to reduced blood flow and/or hyperthermia can result in a vicious cycle that promotes greater intestinal barrier dysfunction through production of these mediators.

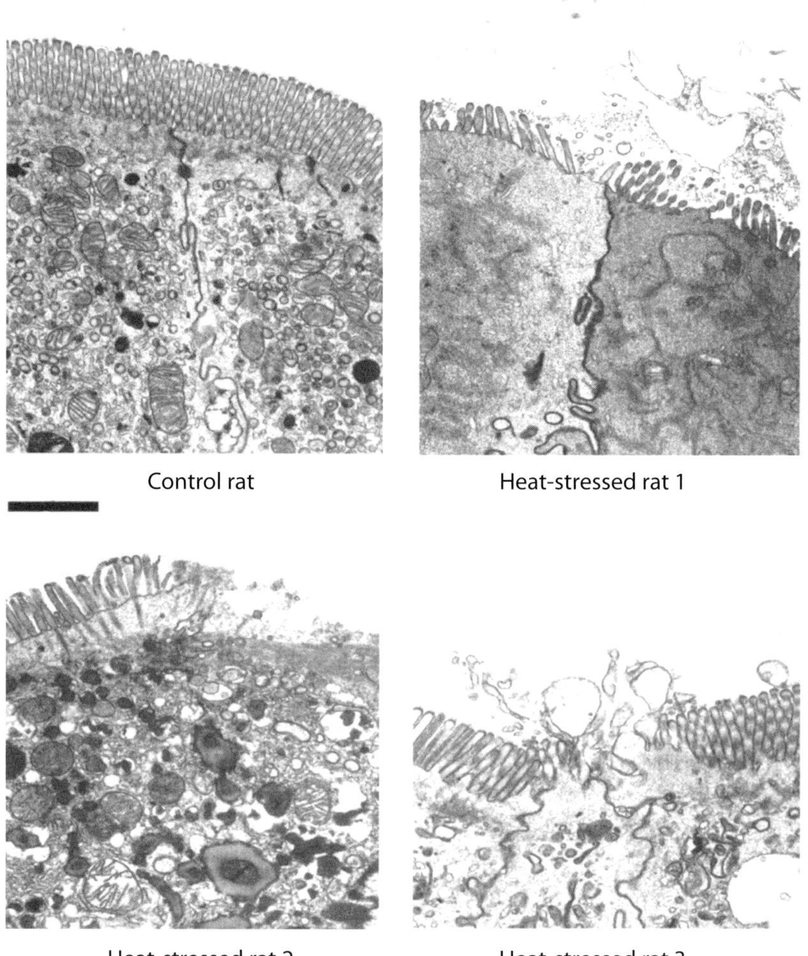

Fig. 4. Transmission electron micrographs of small intestinal epithelial cells from anesthetized control (37°C) and heat-stressed (41.5–42.5°C) rats. Heat stress results in disruption of microvilli and cell membranes, mitochondrial swelling, and cellular vacuolization. Bar = 1 μm. Reprinted from Lambert et al. [11], with permission. Copyright 2002 American Physiological Society.

Nonsteroidal Anti-Inflammatory Drugs (NSAIDS)

NSAIDs are well known to promote GI barrier dysfunction. When used prior to prolonged running, Lambert et al. [25] found that both aspirin and ibuprofen significantly increase GI permeability compared to placebo. Because NSAID use is highly prevalent among athletes, these drugs may play a significant role in the intestinal barrier dysfunction and endotoxemia observed in many individuals during exercise-heat stress.

Effects of Endotoxemia during Exercise-Heat Stress

Endotoxemia in resting humans produces GI symptoms, increased pro-inflammatory cytokine release, and fever. Similar effects are also likely mediated by endotoxemia during exercise-heat stress. Such effects during exercise-heat stress will be discussed below.

GI Symptoms
In the study by Lambert et al. [7] in which increased intestinal permeability was observed following a long-distance triathlon, 61% of the subjects also experienced nausea. Similarly, in the study by Jeukendrup et al. [8], in which plasma LPS and pro-inflammatory cytokine concentrations were significantly elevated also following a long-distance triathlon, 21% of the subjects reported vomiting, with two subjects unable to finish the race due to multiple GI problems (i.e. diarrhea, cramping, vomiting, nausea). Furthermore, in the study by Brock-Utne et al. [2] 80.6% of the runners with high LPS values reported nausea, vomiting and/or diarrhea.

Pro-Inflammatory Cytokine Release
The effects of endotoxemia are mediated through the release of pro-inflammatory cytokines such as interleukin-1, interleukin-6 and tumor necrosis factor-α [8]. If during exercise-heat stress, the concentrations of these cytokines becomes high enough, their effects can lead to greater heat storage (i.e. pyrogenic effect), systemic inflammation, and possible multiple organ damage. The latter effects are likely responsible for the lethality of exertional heat stroke [1]. Figure 5 indicates the possible sequence of events that can occur if intestinal barrier dysfunction and endotoxemia become severe during exercise-heat stress.

In support of the contention that endotoxemia and the release of pro-inflammatory cytokines are responsible for the etiology of heat stroke, it has been shown that administration of gut-cleansing antibiotics [4], anti-LPS hyperimmune plasma [26], and/or corticosteroids [27, 28] prevents heat stress-induced endotoxemia [4, 27] and improves heat stroke survival [26, 28].

Fever
It has been debated as to whether endotoxemia creates a pyrogenic effect (i.e. fever) that increases the rate of rise in core temperature during exercise-heat stress. The evidence is clear that 'endogenous pyrogens' (e.g. interleukin-1, interleukin-6, tumor necrosis factor-α) are released during strenuous exercise [8] and heat stress [5]. Basically, the hypothesis states that endotoxemia and the subsequent release of endogenous pyrogen(s) raises the hypothalamic set point during exercise, promoting a greater rate of heat storage. In support of this hypothesis, Hales and Nagai [29] have demonstrated that the rate of rise in core temperature is greater in heat-stressed

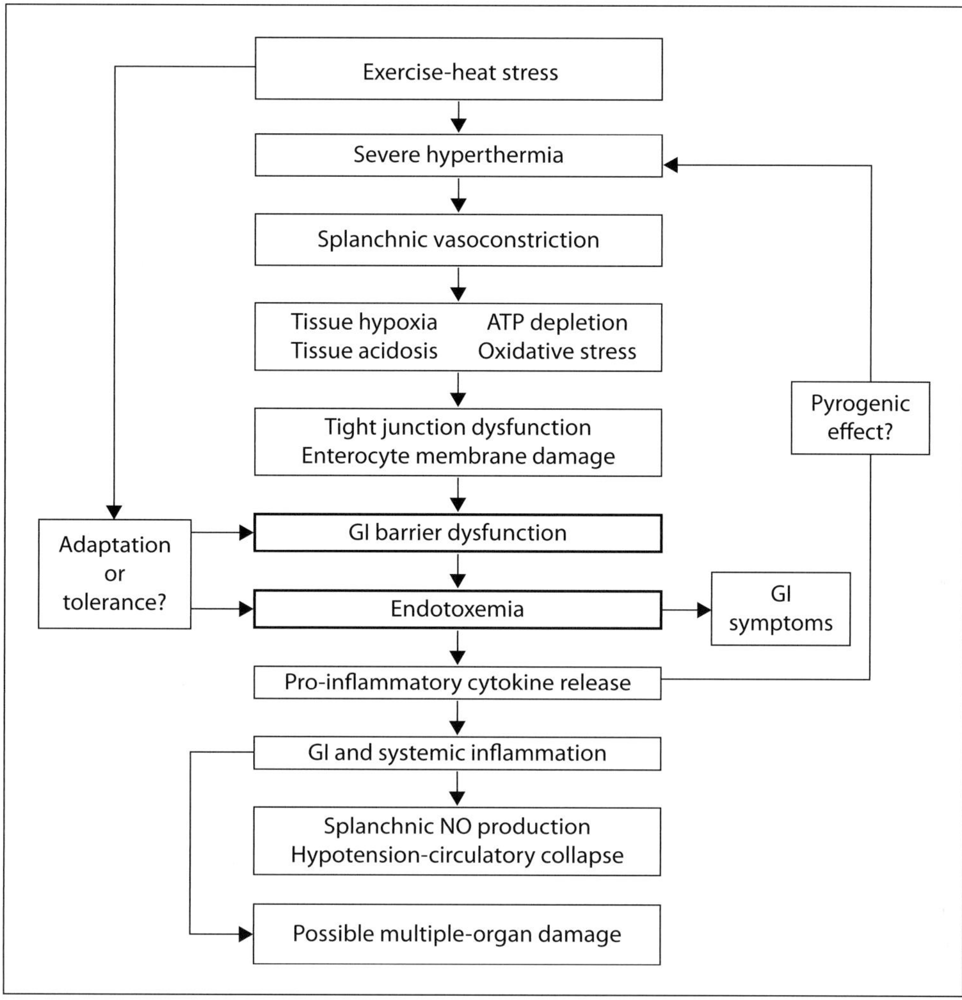

Fig. 5. Flow of events linking exercise-heat stress, intestinal barrier dysfunction, and endotoxemia to possible problems encountered during exertional heat stroke.

rabbits administered LPS compared to control rabbits, but is lower in heat-stressed rabbits administered indomethacin prior to heat stress. This was further supported by Sakurada and Hales [30] who studied trained versus untrained sheep exercising in the heat. In that study, it was found that the rate of rise in core temperature was attenuated when indomethacin was administered prior to the experiments. These authors attributed the effect of indomethacin to the inhibition of prostaglandin E_2 (PGE_2) synthesis caused by LPS-induced release of endogenous pyrogens. It is believed that PGE_2 is the actual mediator for the increase the hypothalamic set-point during LPS-induced fever. Accordingly, Bradford et al. [31] have recently shown that NSAID-induced inhibition of cyclooxygenase-2 lowers core temperatures in humans exercising in the heat, indicating there is an endogenous pyrogen-mediated PGE_2 effect on thermoregulation during exercise-heat stress. In contrast to such findings, Caputa et al. [32] reported that neither LPS nor indomethacin affected thermoregulation when administered to rats already in a hyperthermic steady state (core temperatures, 40–41°C). It should be noted with respect to studies such as those previously mentioned that sensitivity to LPS varies among species, with humans being among the most sensitive and rats among the least sensitive.

The 'Canary in the Coal Mine' during Exercise Heat Stress?

Based on the previously described evidence, it appears GI symptoms, pro-inflammatory cytokine release, and fever-like effects are associated with intestinal barrier dysfunction and endotoxemia during exercise-heat stress. Furthermore, since these problems appear to occur prior to serious complications in many individuals, it is possible that they may serve as 'warning signs' of more severe conditions (e.g. exertional heat stroke) if left unheeded. More studies in this area are needed as no studies to date have specifically examined this hypothesis.

Adaptation to Intestinal Barrier Dysfunction and Endotoxemia

Not all individuals with high core body temperatures develop symptomatic endotoxemia during exercise-heat stress. This is evident in highly trained endurance athletes who expose themselves to hyperthermia and reduced intestinal blood flow on a regular basis in training and competition, yet exhibit no ill effects. Such an observation leads to the question 'Is it possible that either the intestinal barrier adapts to exercise-heat stress or the body acquires a tolerance to gut-derived LPS?' The answers appear to be yes.

Exposure to heat stress appears protective to the intestinal barrier. Moseley et al. [9] found in vitro (i.e. MDCK cells) that a conditioning heat stress (42°C for 90 min) increased the temperature at which tight junction dysfunction occurred and also improved cell survival to a lethal heat stress. These beneficial effects were associated with increased heat shock protein (HSP) accumulation in the cells. Increased expression of HSPs has also been shown in the gut with lower levels of heat stress [33]; however, it appears that the protective effect is not maintained indefinitely, and may be gone within 96 h [9]. Thus, regular exposure to heat stress via exercise training in which the core temperature is raised significantly, but not to dangerous levels, is likely necessary to maintain such protection.

Another adaptation with exercise training that is likely protective to the intestinal barrier is the ability to maintain higher intestinal blood flow at any given level of exercise-heat stress. This was observed in the study by Sakurada and Hales [30] in which greater intestinal blood flow in trained sheep appeared to protect the intestinal barrier and reduce exposure to LPS leading to increased heat tolerance. As previously discussed, reduced GI blood flow likely mediates many of the mechanisms responsible for intestinal barrier dysfunction.

In terms of adaptation to LPS itself, Ryan et al. [34] observed survival of all rats receiving a lethal dose of LPS if they were exposed to heat stress 24 h prior to receiving the LPS. This was compared to 29% survival of rats that were not exposed to heat stress prior to LPS administration. Such results indicate that even an acute heat stress provides protection against the effects of endotoxemia. Recent evidence indicates that heat stress-induced tolerance to the effects of LPS may be related to reduced release of inflammatory mediators, possibly through altered cell signaling mechanisms [35].

Conclusions

In summary, intestinal barrier dysfunction is common among individuals engaged in strenuous, prolonged exercise in the heat, and leads to increased intestinal permeability and possibly endotoxemia. The causes of intestinal barrier dysfunction during exercise-heat stress are mainly related to reductions in intestinal blood flow and intestinal hyperthermia, but may also be related to NSAID use. Endotoxemia can result in a severe inflammatory reaction that may promote a pyrogenic effect, hypotension, and multiple-organ damage. GI symptoms during exercise-heat stress are also many times likely related to intestinal barrier dysfunction and endotoxemia and could therefore serve as a 'canary in the coal mine' warning of a possibly more serious condition (e.g. exertional heat stroke) if left unheeded. It appears, however, that prior exposure to exercise-heat stress may provide some protection against such problems.

References

1   Lambert GP: Role of gastrointestinal permeability in exertional heatstroke. Exerc Sport Sci Rev 2004; 32:185–190.

2   Brock-Utne J, Gaffin S, Wells M, Gathiram P, Sohar E, James M, Morrell D, Norman R: Endotoxemia in exhausted runners after a long-distance race. S Afr Med J 1988;73:533–536.

3   Gathiram P, Wells MT, Raidoo D, Brock-Utne JG, Gaffin SL: Portal and systemic plasma lipopolysaccharide concentrations in heat-stressed primates. Circ Shock 1988;25:223–230.

4   Gathiram P, Wells MT, Brock-Utne JG, Wessels BC, Gaffin SL: Prevention of endotoxaemia by non-absorbable antibiotics in heat stress. J Clin Pathol 1987;40:1364–1368.

5 Bouchama A, Parhar RS, El-Yazigi A, Sheth K, Al-Sedairy S: Endotoxemia and release of tumor necrosis factor and interleukin 1-alpha in acute heatstroke. J Appl Physiol 1991;70:2640–2644.

6 Pals KL, Chang RT, Ryan AJ, Gisolfi CV: Effect of running intensity on intestinal permeability. J Appl Physiol 1997;82:571–576.

7 Lambert GP, Murray R, Eddy D, Scott W, Laird R, Gisolfi CV: Intestinal permeability following the 1998 Ironman triathlon. Med Sci Sports Exerc 1999;31:S318.

8 Jeukendrup AE, Vet-Joop K, Sturk A, Stegen JH, Senden J, Saris WH, Wagenmakers AJ: Relationship between gastro-intestinal complaints and endotoxemia, cytokine release and the acute-phase reaction during and after a long-distance triathlon in highly trained men. Clin Sci (Lond) 2000;98:47–55.

9 Moseley PL, Gapen C, Wallen ES, Walter ME, Peterson MW: Thermal stress induces epithelial permeability. Am J Physiol Cell Physiol 1994;267:C425–C434.

10 Hall DM, Buettner GR, Oberley LW, Xu L, Matthes RD, Gisolfi CV: Mechanisms of circulatory and intestinal barrier dysfunction during whole body hyperthermia. Am J Physiol Heart Circ Physiol 2001;280:H509–H521.

11 Lambert GP, Gisolfi CV, Berg DJ, Moseley PL, Oberley LW, Kregel KC: Selected contribution: hyperthermia-induced intestinal permeability and the role of oxidative and nitrosative stress. J Appl Physiol 2002;92:1750–1761.

12 Dokladny K, Moseley PL, Ma TY: Physiologically relevant increase in temperature causes an increase in intestinal epithelial tight junction permeability. Am J Physiol Gastrointest Liver Physiol 2006;290:G204–G212.

13 Rowell LB: Human cardiovascular adjustments to exercise and thermal stress. Physiol Rev 1974;54:75–159.

14 Hall DM, Baumgardner KR, Oberley TD, Gisolfi CV: Splanchnic tissues undergo hypoxic stress during whole body hyperthermia. Am J Physiol 1999;276:G1195–G1203.

15 Unno N, Menconi MJ, Salzman AL, Smith MSH, Ge Y, Ezzell RM, Fink MP: Hyperpermeability and ATP depletion induced by chronic hypoxia or glycolytic inhibition in Caco-2BBe monolayers. Am J Physiol 1996;270:G1010–G1021.

16 Salzman AL, Wang H, Wollert PS, Vandermeer TJ, Compton CC, Denenberg AG, Fink MP: Endotoxin-induced ileal mucosal hyperpermeability in pigs: role of tissue acidosis. Am J Physiol 1994;266:G633–G646.

17 Parks DA, Bulkley GB, Granger DN: Role of oxygen-derived free radicals in digestive tract diseases. Surgery 1983:415–422.

18 Hall DM, Buettner GR, Matthes RD, Gisolfi CV: Hyperthermia stimulates nitric oxide formation: electron paramagnetic resonance detection of ˙NO-heme in blood. J Appl Physiol 1994;77:548–553.

19 Mercer DW, Smith GS, Cross JM, Russell DH, Chang L, Cacioppo J: Effects of lipopolysaccharide on intestinal injury: potential role of nitric oxide and lipid peroxidation. J Surg Res 1996;63:185–192.

20 Kregel KC, Wall PT, Gisolfi CV: Peripheral vascular responses to hyperthermia in the rat. J Appl Physiol 1988;64:2582–2588.

21 Rowell LB, Brengelmann GL, Blackmon JR, Twiss RD, Kusami F: Splanchnic blood flow and metabolism in heat-stressed man. J Appl Physiol 1968;24:475–484.

22 Horowitz M, Samueloff S: Cardiac output distribution in thermally dehydrated rodents. Am J Physiol 1988;254:R109–R116.

23 Lambert GP, Lang JA, Bull AJ, Pfeifer PC, Eckerson JM, Moore GA, Lanspa SJ, O'Brien JJ: Fluid restriction increases GI permeability during running. Int J Sports Med 2008;29:194–198.

24 Baumgart D, Dignass A: Intestinal barrier function. Curr Opin Clin Nutr Metab Care 2002;5:685–694.

25 Lambert GP, Boylan MW, Laventure J-P, Bull AJ, Lanspa SJ: Effect of aspirin and ibuprofen on GI permeability during exercise. Int J Sports Med 2007;28:722–726.

26 Gathiram P, Wells MT, Brock-Utne JG, Gaffin SL: Antilipopolysaccharide improves survival in primates subjected to heat stroke. Circ Shock 1987;23:157–164.

27 Gathiram P, Gaffin S, Brock-Utne J, Wells M: Prophylactic corticosteroid suppresses endotoxemia in heat-stressed primates. Aviat Space Environ Med 1988;59:142–145.

28 Gathiram P, Wells M, Brock-Utne J, Gaffin S: Prophylactic corticosteroid increases survival in experimental heat stroke primates. Aviat Space Environ Med 1988;59:352–355.

29 Hales JRS, Nagai M: Endotoxaemia is normally a limiting factor in heat tolerance; in: Thermal Balance in Health and Disease: Recent Basic Research. Basel, Birkhauser, 1994, pp 369–372.

30 Sakurada S, Hales JRS: A role for gastrointestinal endotoxins in enhancement of heat tolerance by physical fitness. J Appl Physiol 1998;84:207–214.

Lambert

31 Bradford C, Cotter J, Thornburn M, Walker R, Gerrard D: Exercise can be pyrogenic in humans. Am J Physiol Regul Integr Comp Physiol 2007;292: R143–R149.

32 Caputa M, Dokladny K, Kurowicka B: Endotoxemia does not limit heat tolerance in rats: the role of plasma lipoproteins. Eur J Appl Physiol 2000;82: 142–150.

33 Ruell P, Hoffman K, Chow C, Thompson M: Effect of temperature and duration of hyperthermia on HSP72 induction in rat tissues. Mol Cell Biochem 2004;267:187–194.

34 Ryan AJ, Flanagan SW, Moseley PL, Gisolfi CV: Acute heat stress protects rats from endotoxic shock. J Appl Physiol 1992;73:1517–1522.

35 Sanlorenzo L, Zhao B, Spight D, Denenberg A, Page K, Wong H, Shanley T: Heat shock inhibition of lipolysaccharide-mediated tumor necrosis factor expression is associated with nuclear induction of MKP-1 and inhibition of mitogen-activated protein kinase activation. Crit Care Med 2004;32: 2284–2292.

G. Patrick Lambert, PhD
Department of Exercise Science, Creighton University
2500 California Plaza
Omaha, NE 68178 (USA)
Tel. +1 402 280 2420, Fax +1 402 280 4732, E-Mail plambert@creighton.edu

Marino FE (ed): Thermoregulation and Human Performance. Physiological and Biological Aspects.
Med Sport Sci. Basel, Karger, 2008, vol 53, pp 74–88

Effects of Peripheral Cooling on Characteristics of Local Muscle

Eric Drinkwater

School of Human Movement Studies and Exercise and Sports Science Laboratories,
Charles Sturt University, Bathurst, NSW, Australia

Abstract

While humans maintain body core temperature within a strict homeostatic range, skin and peripheral muscle temperature may experience a wide temperature variation. Much of the literature investigating cooling on human performance involves cooling of the core, though many performance effects relate to cooling of the periphery. No standard method exits to investigate the effects of cooling, so protocols range across a variety of temperatures (10–42°C), temperature assessment methods (skin, intramuscular), cooling mediums (air, water immersion), muscle fibre type (species, fast or slow twitch), contraction type (evoked or voluntary, isometric or dynamic), and isolated versus intact fibres. Despite these variables, there is general agreement that rate properties are slowed with almost any level of cooling thereby most substantially reducing muscle power. The slowed enzymatic processes and slowed nerve conduction that impair rate of force development also likely reduce local muscular endurance during dynamic contractions and impair manual dexterity ($<35°C$). Both the voluntary and evoked force development capacities of muscle is unimpaired until cooling is quite severe ($<27°C$). While most of these effects occur independently of central activation, purposeful core cooling for the purpose of improving athletic performance should be used cautiously to avoid the deleterious effects of peripheral cooling.

Early research investigating work capacity in cold environments centred around the systemic effects of cold for the purposes of survival, often relating to pilots and mariners during the Second World War or other forms of accidental cold water immersion. However manual labour, athletics, and recreation are now occurring in increasingly cold environments so investigating the capacity to perform work and exercise in cold environments beyond mere survival has become necessary. Some forms of sports (e.g. hiking) may involve long periods of exposure to a cold, wet environment where the person is far from constructed shelter, so the threat of central hypothermia is real. While it is possible for individuals to experience central cooling in the outdoor sporting environment and it is important not to downplay the importance of safety [1], it is more common for military personnel, commercial SCUBA divers, skiers, etc. to experience peripheral cooling while the body's core temperature remains normo- or even hyperthermic. Additionally, participants in cold weather activities are usually aware of the discomfort and threat of central cooling and thus insulate their body core, though will often keep their arms and legs exposed (e.g. vests and shorts) to attenuate the rise in core temperature [2]. Therefore, considering recent publications on exercise in the cold as it relates to central cooling [3] and those outlining safety concerns and recommendations for exercise in the cold [1], matters of core cooling and safety (e.g. hypothermia, frost-bite) will not be reiterated here. Also to be excluded will be the use of cooling for the purpose of prolonging exercise in the heat (i.e. pre-cooling) and cold water immersion for the purposes of recovery from exercise which will be addressed in the chapter by Duffield [this vol., pp. 89–103].

Less commonly considered are the performance effects of cooling of the skin [4] and muscles of the limbs [5]. As humans protect the temperature of their body core within a strict homeostatic range, often at the expense of distal temperatures, the effects of muscle cooling occur independent of core temperature [5]. When compared to the core, peripheral muscle experiences a large thermal range on a daily basis, reaching as high as 41°C after exhaustive exercise or below 23°C depending on activity level and environmental temperature [6]. To help understand the effects of cooling on performance characteristics such as muscular strength, power, and endurance, it is necessary to first detail the effects of cooling on the mechanistic properties of muscle such as evoked contractile properties. Numerous experiments have shown in human and other animal models that properties altered by decreasing temperature include voluntary and evoked contractile properties of muscle such as voluntary force, electromyograph (EMG) activity, rate processes such as time to twitch or voluntary contraction or relaxation of a muscle contraction and rate of onset of fatigue (for review, see Bennett [7]). From studying these properties, we can derive greater understanding of the effects of local muscle cooling on human performance.

Factors Effecting Research Findings

Confusion often results when reviewing large amounts of literature on the effects of temperature changes due to the wide variety in data collection methods of data collection and reporting. For instance, when comparing results from different studies, a reader should consider where the temperature was measured, and what a 'normal' temperature is for that site. Under strict homeostatic control, core temperature measured 5–8 cm rectally usually rests at 37°C. Resting intramuscular temperature of the major limb muscles should be approximately 35°C when measured to a depth of 20–30 mm under the skin under thermoneutral conditions. Skin temperature is the most sensitive to ambient temperature and, at approximately thermoneutral ambient temperature of 28°C, skin temperature can be expected to be approximately 33°C. Considering that differences of up to 10°C exist between skin and muscle temperatures, it can be difficult to extrapolate the extent of muscle cooling from the skin temperature, particularly considering individual differences such as limb size and insulation [8]. Finally, some research will only report ambient temperature. With water having 25 times the convection capacity of air [3], knowing if the medium is air or water is also important when comparing ambient temperatures.

A variety of individual differences in study participants and study design can also lead to varying results between studies. Differences such as limb surface area, muscle mass, limb size, and subcutaneous fat will play a substantial role in the extent that a muscle will cool in response to changing ambient temperature [8]. Additionally, while acclimation to whole-body cooling is reasonably well documented, a participant's level of acclimation can also affect results in limb cooling studies. Rintamäki et al. [9] found a substantial decrease in finger vascular resistance and a higher finger temperature after 53 days of Antarctic work. Finally, discrepancies when comparing the results of studies can also reflect differences in contraction type, muscle group and muscle fibre type composition. As a result of all of these differences in study design and participants, there are often small differences in findings between studies that, on the surface, appear to be investigating similar concepts.

Neuromuscular Mechanisms

Muscle Fibre Excitation

Rate processes deal with time-dependent features of a muscle. They are useful in studying the speed at which a particular process is accomplished and may be used to explain such characteristics as power development or speed of a response. The dependence of rate processes between two temperatures is referred to as a 'Q_{10}'. A Q_{10} of 1.0 conveys no change or dependence, less than 1.0 is a negative dependence, and greater than 1.0 is a positive dependence. A score of greater than 2.5 or less than 0.5 convey strong dependence [10].

Rate processes of muscle contractions can be studied using voluntary contractions or contractions evoked using external electrical stimulation. Stimuli can be used to evoke contractions individually (i.e. twitch) or in trains (i.e. tetanus), in intact muscle groups or in isolated muscle fibres. Rate properties of both tetanus and twitch include properties such as shortening velocity, rate of force development, time to peak force, and the half relaxation time. There is a clear relationship

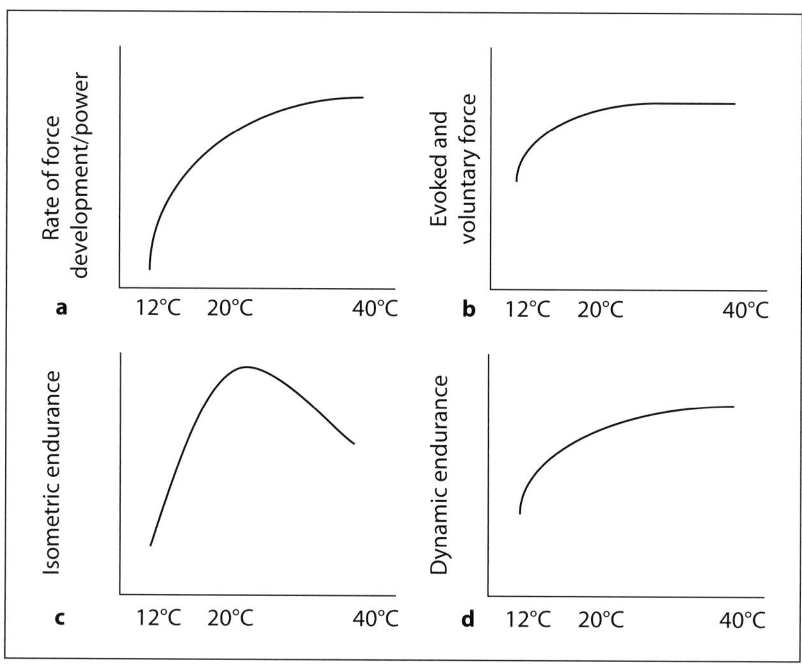

Fig. 1. Schematic representations of skeletal muscle properties in response to cooling: rate of force development/power (**a**), evoked and voluntary force (**b**), isometric endurance (**c**), and dynamic endurance (**d**). Figures drawn from data obtained from a number of studies as discussed in the text.

between rate processes and temperature over physiological temperature ranges (Q_{10} = 1.6–3) (fig. 1a) [10]. The precise Q_{10} varies across species, muscle fibre types, levels of cooling, and specimen preparation (i.e. isolated fibres or intact muscle), as well as the property being studied (rate of force development, relaxation time, etc.), and the type of contraction (i.e. isometric, dynamic) [7]. What can be agreed upon is that almost any level of muscle cooling slows rate processes [11–17]. Faulkner et al. [18] illustrated that the rate of force development of isolated rat muscle was slowed by 22% in cooling to 25°C, while Drinkwater and Behm [17] showed that the rate of isometric twitch and tetanic force development declined by a similar 50 and 46%, respectively, when the plantar flexors were cooled to 22°C. Furthermore, Ranatunga et al. [14] found that time to peak twitch almost doubled in cooling the skin to 12.5°C. The thermal dependence of rate processes also continues into hyperthermic temperature ranges as well, since, as temperature increases above normal muscle temperature, the speed of rate processes continue to increase [13, 19]. The thermal dependence of rate processes should not be considered linear, however. The Q_{10} for rate processes as a whole below 20°C were considerably higher than those obtained for temperatures above 25°C, being 2.75 and 1.48, respectively [14, 16] (fig. 1a).

There is some disagreement on the thermal dependence of different muscle groups. Maximal shortening velocity has been shown to have a similar thermal dependence in both slow- and fast-twitch muscle fibre types when cooled from 35 to 25°C (Q_{10} = 1.8 in fast and 2.0 in slow), though this gap widened when below 20°C (Q_{10} = 2.4 in fast and 3.5 in slow) [20]. However, fast-twitch fibres have also shown a greater thermal dependence (Q_{10} of 2.1–2.2) than in slow twitch (Q_{10} of 1.4–1.6) for shortening velocity [21]. It should be noted that while variation exists between studies, variance is not great and likely results from different methodologies or individual variations among samples. Since differences already exist between fast and slow motor units in properties such as firing frequency of the motor unit and rate of myosin ATPase activity, properties likely impaired by cooling, it seems logical that differences would exist between the different types of motor units.

A variety of factors likely act in concert to slow contraction speed. Changes in rate processes with decreasing temperature may be a consequence of the effect of temperature on metabolic rate [18, 22] and the maximal rate of adenosine triphosphate (ATP) hydrolysis (e.g. actomyosin ATPase) [23]. There is also impairment of calcium release from the sarcoplasmic reticulum (SR) [24], reduced calcium sensitivity [25], and/or impaired kinetics of the muscle fibre action potentials [23, 26, 27] may all play a role in impairing contraction speed. Any or all of these would result in a net reduction in the rate of cross bridges cycling [16] and thus result in slowing the contraction velocity.

Another characteristic effect of muscle cooling is the prolongation of the half relaxation time of the twitch (1/2 RT). The Q_{10} of the 1/2 RT has been shown to be 1.7 in human first dorsal interosseus muscle during cooling from 35 to 25°C [14]. Drinkwater and Behm [17] illustrated a 132% increase in the 1/2 RT of isometric twitches while 1/2 RT of tension developed by tetanic stimulation was 119% longer when human plantar flexors were cooled to 22°C. Similarly, at a skin temperature of 12.5°C, Ranatunga et al. [14] illustrated a 200% increase in 1/2 RT. The prolongation of relaxation may be caused by slowed ATP binding to the ATP-binding site on the myosin head thus causing slower detachment of the myosin cross-bridges [18]. Additionally, slowing of SR ATPase would slow the active process of re-uptake of calcium out of the sarcoplasm, thereby also prolonging contraction [17, 24]. The pronounced increase of 1/2 RT finalizes the high thermal sensitivity of rate processes with muscle cooling.

Voluntary EMG and Power Spectrum
Increasing muscle tension occurs by recruiting additional motor units (i.e. increasing recruitment) or by increasing firing frequency of an already firing

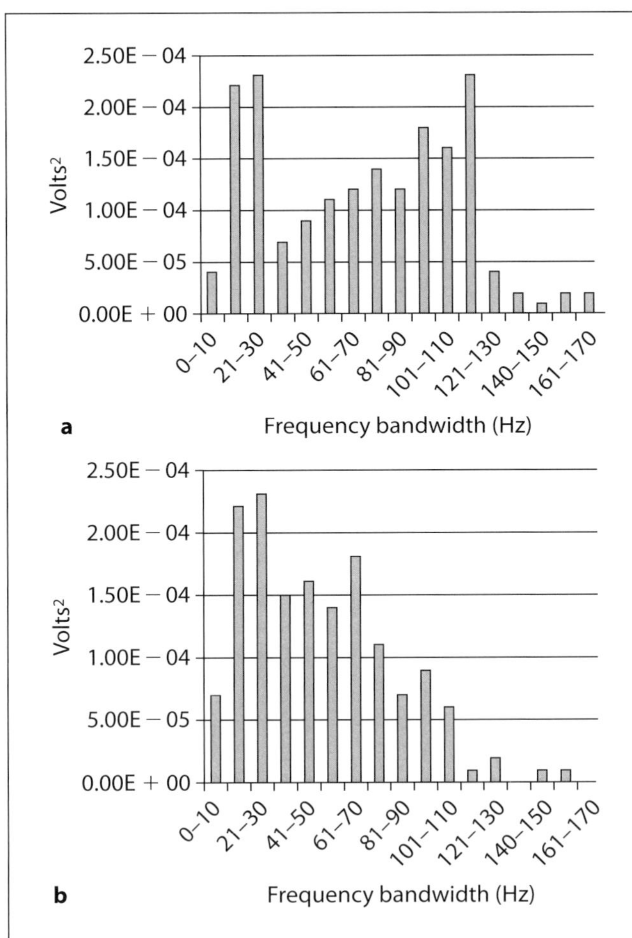

Fig. 2. Frequency distributions of muscle firing when warm (**a**), and hypothermic (**b**). A similar phenomenon occurs when the muscle is non-fatigued (**a**) and fatigued (**b**). Note the shift in power to the lower bandwidths when moving from (**a**) to (**b**). Figures drawn from data obtained from a number of studies as discussed in the text.

motor unit (i.e. increasing rate coding). Since EMG represents the total electrical activity in a given muscle [28], by increasing either or both recruitment or rate coding, the EMG signal will increase. To analyse EMG, the signal is typically rectified and then measured for the area under the curve as integrated EMG (iEMG). Engineering analysis on signals may also involve power spectrum analysis (PSA) [29] which represents the average distribution of the power, or area under the curve, across the frequency range of interest, thereby assessing power distribution as a function of firing frequency (fig. 2). The PSA can be used to analyse iEMG signals and isolate how a certain firing frequency contributed to the total contraction thereby helping to elucidate changes in firing frequency of a muscle and make inferences about the active fibre types. Both the amplitude of EMG and total power of the EMG power spectrum have a linear relationship to increasing force.

Generally, cooling muscle to approximately 20°C increases the iEMG signal during both maximal [17] and submaximal [28] voluntary contractions, as well as isometric [17] and dynamic contraction types. Drinkwater and Behm [17] demonstrated this increase to be 37% in soleus iEMG when cooled to 22°C during maximal seated plantare flexion. By cooling only one leg but measuring iEMG of both legs, Rissenan et al. [2] illustrated that the observed increase in muscle EMG occurred only in the cooled leg, illustrating the effects are likely peripherally mediated. Once muscle temperature decreases below 20°C, however, EMG amplitude rapidly declines.

Possible explanations for this increase in EMG relate to recruiting more (higher threshold) motor units in response to the cold [10]. While muscle force generating ability remains reasonably constant with cooling to approximately 25°C [7, 8, 14, 17, 30], temporal characteristics, as previously discussed, steadily decline thereby reducing the power the muscle fibre can generate [18]. As a result, in an attempt to generate equivalent power at lower temperatures, there appears to be a compensatory mechanism that recruits higher threshold motor units earlier in the recruitment order, referred to as 'compression of the recruitment order' [10]. The muscle fibres continue to be recruited in the same order at lower temperatures (i.e. slow twitch fibres are recruited before fast twitch), but the lower power output of slow twitch fibres at lower temperature causes the faster, more powerful fibres to be recruited earlier. Therefore, at a given workload, a greater number of fast twitch motor units are used at lower temperatures, thereby increasing the EMG for a given workload [17, 28, 31]. There is also some evidence to suggest that recruitment order does not simply compress but there is some change in the recruitment order itself, such that some high threshold motor units are recruited before low threshold motor units as a result of noxious cutaneous stimulation [4]. Regardless, there is a recruitment of higher threshold motor units to achieve the same power output. Having more high threshold motor units active may be expected to lower endurance at that work load due to a higher energy expenditure and greater use of glycolysis and greater lactate production [23]. As a result, it initially appears likely that cooling should reduce endurance capacity of a muscle. While this is the case for dynamic contractions [18, 22, 32], optimal endurance for isometric muscle contractions appease to be at temperatures of 22 to 32°C [6, 8, 23, 28, 33].

Changes in EMG PSA due to decreasing muscle temperature resemble the effects of fatigue-induced muscle wisdom [28]. Muscle wisdom [34] is a phenomenon in which, as muscle fatigue increases, central neural drive attempts to maintain force generation by decreasing the discharge rate so that the discharge rate is maintained at the minimum level in order to maintain the desired force (fig. 2). Muscle wisdom therefore optimizes firing frequency to the desired force output in an attempt to postpone fatigue. Cold muscle shows a similar change in frequency to lower frequency components (fig. 2) [28, 35]. Therefore, while more fibres are active, they are active at a lower frequency [36, 37]. Oska et al. [37] found during the shortening phase of maximal rebound jumps, the mean power spectrum of agonists declined from 124 Hz at 27°C ambient temperature to 82 Hz at 10°C. As a result, a comparison of EMG to force maintenance appears to be optimal at 27°C [8]. Thus, with moderate cooling, fibres are recruited at a lower frequency of discharge [28]. Since low stimulations rates are required to maintain a submaximal force when the muscle is cooled, this lower frequency of discharge could be considered more economical [6]. To be discussed with the functional consequences of local muscle cooling, the performance effect of this maybe limited as the theory of local muscle cooling, since extending endurance does not appear to apply to dynamic contractions [18, 22, 32].

The motor units' ability to tetanize at lower frequencies [6] relates to the prolonged duration of contractions (i.e. 1/2 RT) and prolonged compound muscle action potentials (i.e. M-wave) at cooler temperatures [17]. Since fast twitch fibres are recruited earlier (i.e. compression or reversal of the recruitment order), thus increasing the rate of fatigue, defence mechanisms to earlier fatigue (i.e. optimized firing frequency from prolonged contractions) will be beneficial to fatigue resistance. Therefore, as to be discussed in more detail later, isometric EMG to force maintenance appears to be optimal at 27°C [8]. It seems likely that the substantial variation that exists in muscle temperatures, ranging from 34°C under thermoneutral ambient temperature to 22°C and below in situ [17], mean that the muscle has made adaptations to allow muscle contractions to continue under less than thermoneutral conditions. Why this only seems to apply to isometric conditions and not dynamic work is not clear.

Functional Consequences of Cold and Fatigue on Muscle Function

Maximal Voluntary Contractions
Similar to the Q_{10}, an R_{10} score is also used to illustrate thermal dependence. Rather than being for rate processes, however, the R_{10} is used to express the thermal dependence of force. At a muscle temperature of 25–35°C, isometric force production has low temperature dependence [8, 14, 17, 30]. Binkhorst et al. [30] found that the thermal sensitivity of voluntary grip strength could be represented by a Q_{10} of 1.2 between 38 and 22°C. Both Drinkwater and Behm [17] and Clarke and Wojciechowics [33] found that as temperature was lowered, neither initial force nor final strength were significantly lower with cold before or after a fatiguing

protocol. Ranatunga et al. [14] actually found an 8% increase in voluntary force with cooling first dorsal interosseus muscle to 25°C [14]. However, once temperatures get very cold, there are clear decrements in voluntary force. Ranatunga et al. [20] did not find any decrement in voluntary force at a muscle temperature of 25°C but found it had decreased by 30% in the 12–15°C range. In cooling finger flexors to below 27°C, Cornwall et al. [11] showed a decrease of isometric grip strength by 14.8% in males and 30.5% decrease in females, though the authors did not offer an explanation of why the sex difference existed. When comparing studies such as Drinkwater and Behm [17] who showed no effect in cooling the soleus to 22°C, and Cornwall et al. [11] who showed a significant effect in cooling finger flexors to 27°C could include the size of the musculature since different muscle groups show differences in maximal activation under normal circumstances [38]. Davies et al. [12] cooled the triceps surae to 24°C and found an 18% decrement in voluntary force, though cooling took 45 min compared to Ranatunga et al. [14] which took only 15 min so differences could also exist due to the rate of cooling. While the temperature at which cooling begins to impair maximal voluntary force is not exact, the general consensus is that muscle temperature above 27°C will not inhibit maximal force [12].

Rather than report muscle temperature, several studies report water, skin, or air temperature, a factor that may also account for varying results. However, such studies are often more interesting from a practical rather than a mechanistic perspective, as they have greater contextual validity to the occupational or performance setting. For example, Imamura et al. [39] found no changes in grip strength despite a decrease in skin temperature to 11°C. However, Holewijn and Heus [35] found that immersion in 15°C water for 30 min resulted in a significant 21.8% decline in maximal force of grip strength. Clarke et al. [8] found that MVC of the finger flexors did not change with no less than 30 min immersion in a water temperature of 18°C, but fell by 40% only once the water temperature was 2°C. Again, when reporting only ambient temperatures, there are a wide variety of factors that lead to each individual participant having different muscle temperatures.

The exact mechanism of changes in voluntary tension during peripheral cooling has not been identified, though it is apparent that most effects are peripherally mediated [2, 40]. Using electrical stimulation of a muscle allows for the study of muscle activation independent of central activation; there is a 30–50% decrement in twitch force with cooling below 24°C [12, 14, 17]. Cooling of the core to 35.8°C but maintaining muscle temperature did not result in any difference in performance in a variety of motor tasks, accounting for 85–98% of the variance in performance in muscle temperature [40]. Rissen et al. [2] also found that the effects of local cooling on EMG were restricted only to the cooled limb. While a small reduction in firing frequency or rate of cross-bridge cycling with mild and moderate cooling may not be sufficient to reduce force output, when muscle temperature becomes very cold, firing frequency and/or rate of cross-bridge cycling may be reduced below the critical level necessary to exert maximal force.

Sub-Maximal Voluntary Contractions (Local Muscular Endurance)
Pamphlets on cold water survival from the Canadian Red Cross Society indicate that a swimmer has only 10% of their warm water swimming capacity in cold water. While the limitations on whole body endurance are well documented [3], it is therefore surprising to find that it has been generally well documented that the thermal dependence of endurance of isometric contractions is best described by a bell shape, usually peaking between temperatures of 22–32°C [6, 8, 23, 28, 33]. If optimal muscular endurance occurs with mild-to-moderate cooling of muscle, why is exercise capacity so limited, considering the lack of central hypothermia at the point of swimming failure [41]. Some differences exist, however, between endurance of isometric contractions versus dynamic work. Oksa et al. [32] demonstrated that a decrease of only 2.6°C of forearm flexors led to a 250% increase in the rate of fatigue. Oksa et al. [32] also demonstrated that the decrement in MVC after a fatigue protocol was more pronounced in mildly cooled muscle (29°C, 17% decrement) than in normo-thermic muscle (34°C, 15% decrement). These two results of Oksa et al. [32] using dynamic contractions are in contrast to those of Drinkwater and Behm [17] who found neither of these using isometric contractions. Bergh and Ekblom [22] also showed that high-intensity (sprinting) endurance seems more susceptible to impairment, decreasing by 55% with cooling from 39 to 30°C. Faulkner et al. [18] also found that sustained power was decreased by 62% when cooling muscle to 25°C.

As previously discussed, compression (or reversal) of the recruitment order recruits high-threshold motor units earlier, thereby increasing EMG, so a lower endurance would be expected. However, as conduction velocity and rate of force development slow and 1/2 RT increases, a lower firing frequency is necessary [6, 26], thus compensating for the increased active muscle mass and thereby allowing increased endurance. Segal et al. [6] found that at a stimulation rate of 28 Hz, the extensor digitorum longus had single twitches at 30–40°C, unfused tetanus at 25°C, and fused tetanus at 20°C. This illustrates that a lower stimulation rate is required to achieve fused tetanus at a cold muscle temperature [42]. This effect is attributed to the accompanying decrease in myosin ATPase activity and calcium sequestering by the SR as muscle temperature decreases [6]. These effects result in a lower firing frequency of the muscle to hold a given force at a given temperature.

Since maximal force output does not have a high thermal sensitivity, but dynamic (rate) properties of muscle, such as power, are very thermally sensitive, it seems intuitive that there is a difference in endurance between static and dynamic contractions. Endurance differences between isometric and dynamic contractions are likely related to differences in the fatigue patterns and ATP utilization between isometric and dynamic contractions under normothermic conditions, with isometric contractions having a lower ATP demand [43]. Also increasing the ATP demand would be the cold-induced increase in muscle stiffness illustrated [44]. Increased stiffness would add resistance to dynamic contractions, increasing ATP demand, thereby reducing endurance.

The other rate process concerning fatigue that has received little research attention is the rate of recovery from fatigue. The rate of recovery from fatigue also appears impaired by cooling [17]. During a high-intensity isometric contraction under normothermic conditions, many evoked properties (e.g. torque, rate of torque development) are potentiated [17], likely at least partially as a result of impaired reuptake of calcium from the sarcoplasm due to impairment of SR ATPase. Slowing of calcium reuptake would maintain the exposure of troponin-C to calcium, thus serving to maintain force without additional stimulation. While investigating the effects of cooling on the recovery from fatigue, Drinkwater and Behm [17] found that the return of these potentiated values back to baseline is impaired by cooling. By further impairing calcium reuptake [24], local muscular cooling not only helps offset what would otherwise be a rapid onset of fatigue due to greater activation of fast twitch muscle fibres, but, by slowing recovery of this potentiation, also serves to prolong endurance.

Power

As previously discussed, there seems to be a difference in the response of dynamic versus isometric contractions, a fact that is likely elicited by the high thermal sensitivity of rate processes. As a result, the muscle characteristic most susceptible to cold-induced decrement is power [5, 7, 13, 18, 21, 30, 37, 42]; the greater the extent of cooling and the greater the power required, the greater the decrement in power [22, 45]. De Ruiter and De Haan [42] found that Q_{10} for isolated adductor pollicis muscle was 2.0 between 37.1 and 31.4°C, and was 6.9 between 25.6 and 22.2°C. With 1 h of whole body exposure to 10°C, time to reach maximal EMG of the arm muscles during throwing was shown by Oksa et al. [45] to increase by 30–42%. Oksa et al. [45] also showed that this level of cooling resulted in a 5.6–9.4% reduction in throwing velocity, the greater decrements occurring with heavier balls [45]. Greater impairment occurred with greater cooling and with higher intensity pedalling.

The high Q_{10} but low R_{10} of muscle likely indicates that the thermal sensitivity of power is linked to the rate of force development rather than the amount of force generated as indicated by the leftward shift of the force-velocity curve [22, 35]. Studying isometric grip strength, Holewijn and Heus [35] showed a 50% decrease in rate of voluntary force development, while Drinkwater and Behm [17] similarly illustrated that rate of isometric voluntary force development declined by 48%. Faulkner et al. [18] accounted for most of the decrements in power to be due to a decrease in shortening velocity though Bergh and Ekblom [22] concluded the opposite; changes in power were a result of force decrements.

Decreases in contractile kinetics can be explained by a decreased cycling frequency. In addition, it has been shown that cooling the muscle may interfere with neuromuscular transmission of the muscle fibres and conduction velocity within the muscles [17, 32] and hence maximum speed thereby setting limits on power and speed [46]. Further reducing power is a decreased sensitivity of the muscle spindles as illustrated by the dose-dependent relationship of drop-jump flight time, average jump force, and jump take-off velocity and temperature [37]. Finally, Sekihara et al. [44] illustrated a cold-induced increase in muscle stiffness that would likely offer greater resistance to high velocity movements. Therefore, activities that involve high power, and particularly those dependent on the stretch-shortening cycle, are very sensitive to cold.

A potential exception may exist regarding the effects of cooling on power output. A variety of noxious stimuli (e.g. cold, electric) have been used to activate cutaneous afferents, resulting in earlier recruitment of high threshold motor units (i.e. reversal of the recruitment order) [4]. While it has not been tested experimentally, the earlier recruitment of high threshold motor units elicited by cold may result in a rightward shift of the force velocity curve if it could be accomplished without cooling of the deep tissues. Therefore, short-duration rapid cooling of the skin could have beneficial effects to very-high-power, short-duration events (e.g. weightlifting, sprinting).

Manual Dexterity

Several studies have illustrated impaired finger dexterity when exposed to sub-zero ambient temperatures [39]. However, these studies illustrated the decrement on non-specific, unpractised tasks. Such decrements may not apply to well-practiced skills [47], particularly when the skill is learned under warm conditions [48]. Impaired accuracy of performance may also not occur when the skill, such as maintaining submaximal force production of the elbow flexors, is of very low complexity or of a larger muscle group [31] though there is demonstrated loss of finger dexterity [39]. Any decrease in coordination that has been observed has been explained to be likely linked to changes in the interaction between the agonist and antagonist muscle groups [45, 49]. This may help explain why well-practiced skills are not affected by muscle cooling: one of the many adaptations to well-rehearsed skills is improved efficiency of the agonist and antagonist muscle groups. After the biceps were exposed to 5°C water for 30 min, Sekihara et al. [44] concluded that there is also a cold-induced reduction in proprioception that can also lead to an increase in positional errors of a limb.

Conclusion

Local muscular cooling commonly occurs in a variety of occupational, sporting, and recreational contexts. While maximal isometric force and endurance are generally unimpaired or even enhanced by moderate cooling, local muscle cooling appears to elicit a decrement in most types of performance where dynamic contraction (i.e. work) is involved including endurance and power. While the impact of cooling on the impairment of coordinating well-practiced physical skills is still limited, there is likely an impairment of proprioception. While much of the existing evidence indicates that force and endurance characteristics of muscle are unimpaired or enhanced by mild cooling, such evidence is generally collected using isometric contractions. When research investigates dynamic contractions, most properties appear to be impaired by even moderate levels of cooling. Therefore, the general recommendation is to avoid local muscular cooling for activities involving dynamic contractions.

References

1 Castellani JW, Young AJ, Ducharme MB, Giesbrecht GG, Glickman E, Sallis RE: American College of Sports Medicine position stand: prevention of cold injuries during exercise. Med Sci Sports Exerc 2006;38:2012–2029.
2 Rissanen S, Oksa J, Rintamäki H, Tokura H: Effects of leg covering in humans on muscle activity and thermal responses in a cool environment. Eur J Appl Physiol 1996;73:163–168.
3 Noakes TD: Exercise and the cold. Ergonomics 2000;43:1461–1479.
4 Yona M: Effects of cold stimulation of human skin on motor unit activity. Jpn J Physiol 1997;47:341–348.
5 Oksa J, Rintamäki H: Dynamic work in cold. Arctic Med Res 1995;54:29–31.
6 Segal SS, Faulkner JA, White TP: Skeletal muscle fatigue in vitro is temperature dependent. J Appl Physiol 1986;61:660–665.
7 Bennett AF: Temperature and muscle. J Exp Biol 1985;115:333–344.
8 Clarke RSJ, Hellon RF, Lind AR: The duration of sustained contractions of the human forearm at different muscle temperatures. J Physiol (Lond) 1958;143:454–473.
9 Rintamäki H, Hassi J, Smolander J, Louhevaara V, Rissanen S, Oksa J, Laapio H: Responses to whole body and finger cooling before and after an Antarctic expedition. Eur J Appl Physiol 1993;67: 380–384.
10 Rome LC: Influence of temperature on muscle recruitment and muscle function in vivo. Am J Physiol Regul Integr Comp Physiol 1990;259: R210–R222.
11 Cornwall MW: Effect of temperature on muscle force and rate of muscle force production in men and women. J Ortho Sports Phys Ther 1994;20: 74–80.
12 Davies CT, Mecrow IK, White MJ: Contractile properties of the human triceps surae with some observations on the effects of temperature and exercise. Eur J Appl Physiol 1982;49:255–269.
13 Davies CT, Young K: Effect of temperature on the contractile properties and muscle power of triceps surae in humans. J Appl Physiol 1983;55:191–195.
14 Ranatunga KW, Sharpe B, Turnbull B: Contractions of a human skeletal muscle at different temperatures. J Physiol (Lond) 1987;390:383–395.
15 Ranatunga KW, Wylie SR: Temperature effects on mammalian contraction. Biomed Biochim Acta 1989;48:S530–S535.
16 Elmubarak MH, Ranatunga KW: Temperature sensitivity of tension development in a fast-twitch muscle of the rat. Muscle Nerve 1984;7:298–303.
17 Drinkwater EJ, Behm DG: Effects of 22°C muscle temperature on voluntary and evoked muscle properties during and after high intensity exercise. Appl Physiol Nut Metab 2007;32:1043–1051.

18 Faulkner JA, Zerba E, Brooks SV: Muscle temperature of mammals: cooling impairs most functional properties. Am J Physiol Regul Integr Comp Physiol 1990;259:R259–R265.

19 Ball D, Burrows C, Sargeant AJ: Human power output during repeated sprint cycle exercise: the influence of thermal stress. Eur J Appl Physiol 1999; 79:360–366.

20 Ranatunga KW: The force-velocity relation of rat fast- and slow-twitch muscles examined at different temperatures. J Physiol (Lond) 1984;351:517–529.

21 Bennett AF: Thermal dependence of muscle function. Am J Physiol Regul Integr Comp Physiol 1984;247:R217–R229.

22 Bergh U, Ekblom B: Influence of muscle temperature on maximal muscle strength and power output in human skeletal muscles. Acta Physiol Scand 1979;107:332–337.

23 Edwards RH, Harris RC, Hultman E, Kaijser L, Koh D, Nordesjo LO: Effect of temperature on muscle energy metabolism and endurance during successive isometric contractions, sustained to fatigue, of the quadriceps muscle in man. J Physiol (Lond) 1972;220:335–352.

24 Kössler F, Kuchler G: Contractile properties of fast and slow twitch muscles of the rat at temperatures between 6 and 42°C. Biomed Biochem Acta 1987; 46:815–821.

25 Sweitzer NK, Moss RL: The effect of altered temperature on Ca2(+)-sensitive force in permeabilized myocardium and skeletal muscle: evidence for force dependence of thin filament activation. J Gen Physiol 1990;96:1221–1245.

26 Segal SS, Faulkner JA: Temperature-dependent physiological stability of rat skeletal muscle in vitro. Am J Physiol Cell Physiol 1985;248:C265–C270.

27 Ward MR, Thesleff S: The temperature dependence of action potentials in rat skeletal muscle fibres. Acta Physiol Scand 1974;91:574–576.

28 Petrofsky JS, Lind AR: The influence of temperature on the amplitude and frequency components of the EMG during brief and sustained isometric contractions. Eur J Appl Physiol 1980;44:189–200.

29 Kwatny E, Thomas DH, Kwatny HG: An application of signal processing techniques to the study of myoelectric signals. IEEE Trans Biomed Eng 1970; 17:303–313.

30 Binkhorst RA, Hoofd L, Vissers AC: Temperature and force-velocity relationship of human muscles. J Appl Physiol 1977;42:471–475.

31 Meigal AY, Oksa J, Gerasimova LI, Hohtola E, Lupandin YV, Rintamäki H: Force control of isometric elbow flexion with visual feedback in cold with and without shivering. Aviat Space Environ Med 2003;74:816–821.

32 Oksa J, Ducharme MB, Rintamäki H: Combined effect of repetitive work and cold on muscle function and fatigue. J Appl Physiol 2002;92:354–361.

33 Clarke DH, Wojciechowics RA: The effect of low environmental temperatures on local muscular fatigue parameters. Am Correct Ther J 1978;32: 35–40.

34 Marsden CD, Meadows JC, Merton PA: 'Muscular wisdom' that minimizes fatigue during prolonged effort in man: peak rates of motoneuron discharge and slowing of discharge during fatigue. Adv Neurol 1983;39:169–211.

35 Holewijn M, Heus R: Effects of temperature on electromyogram and muscle function. Eur J Appl Physiol 1992;65:541–545.

36 Winkel J, Jorgensen K: Significance of skin temperature changes in surface electromyography. Eur J Appl Physiol 1991;63:345–348.

37 Oksa J, Rintamäki H, Rissanen S: Muscle performance and electromyogram activity of the lower leg muscles with different levels of cold exposure. Eur J Appl Physiol 1997;75:484–490.

38 Behm D, Whittle J, Button D, Power K: Intermuscle differences in activation. Muscle Nerve 2002;25: 236–243.

39 Imamura R, Rissanen S, Kinnunen M, Rintamäki H: Manual performance in cold conditions while wearing NBC clothing. Ergonomics 1998;41:1421–1432.

40 Giesbrecht GG, Wu MP, White MD, Johnston CE, Bristow GK: Isolated effects of peripheral arm and central body cooling on arm performance. Aviat Space Environ Med 1995;66:968–975.

41 Wallingford R, Ducharme MB, Pommier E: Factors limiting cold-water swimming distance while wearing personal floatation devices. Eur J Appl Physiol 2000;82:24–29.

42 De Ruiter CJ, De Haan A: Temperature effect on the force/velocity relationship of the fresh and fatigued human adductor pollicis muscle. Pflügers Arch 2000;440:163–170.

43 Kay D, St Clair Gibson A, Mitchell MJ, Lambert MI, Noakes TD: Different neuromuscular recruitment patterns during eccentric, concentric and isometric contractions. J Electromyogr Kinesiol 2000;10:425–431.

44 Sekihara C, Izumizaki M, Yasuda T, Nakajima T, Atsumi T, Homma I: Effect of cooling on thixotropic position-sense error in human biceps muscle. Muscle Nerve 2007;35:781–787.

45 Oksa J, Rintamäki H, Mäkinen T, Hassi JHR: Cooling-induced changes in muscular performance and EMG activity of agonist and antagonist muscles. Aviat Space Environ Med 1995;66:26–31.

46 Ranatunga KW: Changes produced by chronic denervation in the temperature dependent isometric contractile characteristics of rat fast and slow twitch skeletal muscles. J Appl Physiol 1977;273:255–262.

47 Tikuisis P, Keefe AA, Keillor J, Grant S, Johnson RF: Investigation of rifle marksmanship on simulated targets during thermal discomfort. Aviat Space Environ Med 2002;73:1176–1183.

48 Oksa J, Rintamäki H, Mäkinen T: The effect of training of military skills on performance in cold environment. Mil Med 2006;171:757–761.

49 Oksa J, Rintamäki H, Mäkinen T, Martikkala V, Rusko H: EMG-activity and muscular performance of lower leg during stretch-shortening cycle after cooling. Acta Physiol Scand 1996;157:1–8.

Eric Drinkwater, PhD
School of Human Movement Studies and Exercise and Sports
Science Laboratories, Charles Sturt University
Bathurst NSW 2795 (Australia)
Tel. +61 2 6338 6116, Fax +61 2 6338 4065, E-Mail edrinkwater@csu.edu.au

Drinkwater

Marino FE (ed): Thermoregulation and Human Performance. Physiological and Biological Aspects.
Med Sport Sci. Basel, Karger, 2008, vol 53, pp 89–103

Cooling Interventions for the Protection and Recovery of Exercise Performance from Exercise-Induced Heat Stress

Rob Duffield

School of Human Movement Studies and Exercise and Sports Science
Laboratories, Charles Sturt University, Bathurst, NSW, Australia

Abstract
The aim of this chapter is to review the literature on the use of cooling interventions in the protection of
and recovery of performance from exercise-induced heat stress. This chapter will deal primarily with the
effects of pre-cooling on the improvement in exercise performance and the effects of post-exercise cooling on recovery. While pre-cooling has received much research attention, the mechanisms resulting in
enhanced performance remain equivocal and moreover, pre-cooling has previously only been considered effective for endurance performance. More recent research describing the effects of pre-cooling on
exercise performance and prevention of heat-related illness will be examined. This chapter will also deal
with the suppression of exercise performance following heat stress and the use of cooling methods to
improve the recovery of muscle function and subsequent exercise performance. Given the use of cold
water immersion as a recovery practice of many athletes, a surprising lack of research has been conducted on the effects of cooling as a recovery tool from heat stress. As such, this chapter will discuss the
use of cooling interventions on both the prevention of heat stress and recovery of performance from
exercise-induced heat stress.

High internal body temperatures resulting from a disparity between the production and loss of heat during prolonged exercise have been well established as
potentially harmful to physiological functioning and responsible for the reduction
in exercise performance [1, 2]. This exercise-induced rise in internal thermal load
is a normal consequence of any prolonged form of exercise, and is controlled via
the interaction of a series of vascular, muscular and metabolic responses [3].
Accordingly, the control of internal thermal load during exercise in the heat, predominantly via evaporative mechanisms, and concomitant maintenance of central
blood pressure and cardiac supply to working musculature is a demanding challenge

[3]. Given the high proportion of community and athletic activities that occur in warm climates and associated risks to health and performance of exercise-induced thermal stress, strategies to counter or alleviate any imposed heat stress have become popular [4, 5].

A popular intervention used to alleviate the compounding effects of exercise in the heat involves whole- or part-body cooling [4, 5]. Pre-cooling has been of interest from both a physiological [6, 7] and performance [8, 9] perspective for some time, and involves methods that aim to reduce skin and core temperature by the application of cold substances or micro-environments to the body. Despite a reasonable body of literature describing physiological changes and performance improvements with pre-cooling, the actual mechanisms that result in ergogenic benefits in the heat remain equivocal [2, 10]. Further, until recently it seemed pre-cooling was only of benefit to endurance athletes [5, 10]. However, recent research may provide further understanding about why and who pre-cooling may benefit. Of further interest, the increased professionalism and demands of sport have increased the focus on effective recovery and preparation for ensuing bouts of competition or training [11]. The use of cooling interventions as a recovery tool has gained much popularity; however, the recovery from exercise-induced heat stress has remained mainly of medical or therapeutic interest [12] and, further, the mechanisms underlying post-exercise cooling for recovery from exercise remain equivocal [13].

Given recent advances in published research on pre-cooling, field-based data collection on thermoregulatory responses and the use of cold-water immersion as a recovery tool, the use of cryotherapy procedures in both the protection and recovery from exercise-induced perturbations in thermoregulation is of importance. Accordingly, this chapter will discuss two aspects; first the use of pre-cooling procedures for the maintenance and/or enhancement of exercise performance in heat stress and second, the use of post-exercise cooling in the recovery from exercise in the heat.

Pre-Cooling for Protection and Performance Improvement in the Heat

Methods of Pre-Cooling
Given the faster rate of heat storage during exercise in the heat, and associated reduction or termination of exercise [14, 15], the prevention of an increased rate of heat storage is critical. Accordingly, an important concept that is reported in pre-cooling literature is the dose dependency of cooling to create a 'heat sink' to absorb metabolic or environmental heat [16]. A range of pre-cooling methods have been used to reduce skin or core temperature including whole-body (showers, ice-baths, cold rooms) and part-body (vests, sprays and fans, towels) methods. This is an

important issue in the effective implementation of pre-cooling methods, as the extent and duration of the cooling method dictate the effectiveness of resultant physiological responses and ensuing performance benefits [17]. Moreover, the duration of the physiological effects of cooling depend on the length of exposure, with most evidence indicating the effects of pre-cooling lasting 30–40 min [6, 18].

Comparisons within and between studies using whole- versus part-body cooling methods not surprisingly indicate greater physiological perturbations from whole-body cooling [19]; however this may not always necessitate greater performance improvement [20]. Interventions involving whole-body cooling generally result in a 0.3–0.8°C lower core temperature throughout exercise [18, 21], while part-body interventions may often only reduce regional skin temperature [20, 22]. Clements et al. [23] have reported more efficient cooling in cold and ice-water immersions than tepid water, indicating temperature of the micro-environment as important but not the sole determinant of cooling effectiveness. As evidence of this, Duffield and Marino [24] reported that mean skin and core temperatures were blunted only in an ice-bath compared to an ice-vest condition following 15 min of respective cooling. However, part-body cooling can induce significant physiological and performance effects if the intervention is of sufficient duration [9, 22] or includes convective as well as conductive processes [25]. Further, it must be noted that changes in skin temperature, without reductions in core temperature, may still result in a shift in plasma volume [26], although in a normothermic or euhydrated state, this may have minimal affect on performance.

Pre-Cooling and Endurance Performance
Historically, an abundance of pre-cooling research has focused on endurance performance in moderate to hot conditions [4, 5]. However, the range of methods, conditions and exercise protocols used has meant that whilst pre-cooling is accepted as being ergogenic for endurance sports, the mechanisms of performance improvement are somewhat unclear. Early and commonly reported research on pre-cooling for endurance exercise reported cold air or water immersion resulted in lower core temperature and prolonged times to exhaustion or increased rates of work [6, 21, 27]. These studies that applied pre-cooling manoeuvres specifically to influence exercise performance generally indicated that cardiovascular and metabolic factors were the mechanisms through which performances were improved.

Olschewski and Brück [21] reported a 12% improvement in cycling endurance time, with a significant reduction in core temperature (0.3°C) following 2×15 min ambient pre-cooling procedures. Minimal differences were reported between experimental conditions for stroke volume or heart rate values; although a marginally lower cardiac output and small increase in a-vO$_{2 \text{ diff}}$ were highlighted during moderate exercise following pre-cooling, but were not present at maximal intensities. The authors highlighted the reduced cardiovascular strain and increased

O_2 use as potential mechanisms for the improved performance. Similarly, Hessemer et al. [6] also reported an increased O_2 pulse following pre-cooling with reduced core temperatures (0.4°C), decreased sweat rates and increased work in a 1-hour trial. Again, conclusions indicated that performance was improved following pre-cooling due to an increased O_2 supply to the working musculature. In further support, Lee and Haymes [18] concluded that pre-cooling improved running time to exhaustion in the heat due to lower core temperatures and lowered stress on cardiovascular and metabolic systems. The authors concluded that the reduced sweating response and cardiovascular load assists thermoregulatory control and consequently improves O_2 supply. This collection of work exemplifies studies that highlight the proposed reduction in cardiovascular and metabolic strain as mechanisms resulting in exercise performance improvements.

The tone of these highlighted studies is further evident in pre-cooling studies over the ensuing decade, as a collection of research continued to indicate performance improvements with pre-cooling by basing these improvements on mechanisms proposed by the aforementioned studies. Previous reviews on this topic [2, 4, 5] have sufficiently covered the range of studies completed on pre-cooling and endurance performance. A consensus exits amongst the variety of studies that include constant-intensity until exhaustion [18, 22, 28], total maximal work performed in a set time [29, 30] or time to complete variable-paced efforts of set distance [9, 20], that pre-cooling results in improvements in endurance exercise performance.

Despite the obvious consensus of performance improvements following precooling, there is a distinct lack of evidence of the mechanisms resulting in performance improvement. A predominance of pre-cooling literature has related exercise improvements to a reduction in cardiovascular and metabolic strain [9, 18], while recent literature [15, 20] highlights centrally mediated mechanisms of a reduction in motor unit recruitment and activation in response or anticipation of the thermal load [31]. Regardless of the proposed mechanisms, pre-cooling has been reported as only effective in improving prolonged, continuous exercise performance. However, an evaluation of the effects of pre-cooling on other forms of exercise may be of benefit to further understanding that the mechanisms' regulating performance is improvement by pre-cooling.

Pre-Cooling and High-Intensity Exercise Performance
Somewhat counter-intuitively, the proposed benefits for short duration, high-intensity exercise are similar to those described for endurance exercise. It has been proposed that increases in blood volume due to peripheral vasoconstriction and resultant increases in muscle blood flow and oxygen availability may improve short duration, high-intensity exercise performance [32]. However, according to previous muscle function literature, reduced muscle temperatures should decrease

high-intensity performance due to altered mechanical power output of muscle fibres at lower temperatures [33].

Crowley et al. [34] have previously reported an impairment of peak power and work in a 30-second Wingate test following pre-cooling of the lower-body. In contrast, Marsh and Sleivert [32] reported a small increase in mean power output for a 70-second cycle ergometer effort following torso cooling in cold water immersion (12–14°C). While no mechanistic explanations were provided, it seems that reducing core temperature, without affecting the active musculature involved in exercise, may provide performance benefits to high-intensity exercise of shorter durations. Finally, Sleivert et al. [35] reported that while pre-cooling the torso had a minimal effect, cooling the torso and legs reduced peak and mean 45-second power (3–8%). The authors further indicated that a short warm-up prior to exercise ameliorated these noted declines from pre-cooling, concluding that pre-cooling impairs contractile function or anaerobic metabolism. Accordingly, it is evident that given the small volumes of available literature on pre-cooling for short-duration, high-intensity exercise, clear performance benefits are somewhat lacking and in fact, these procedures may be detrimental for forms of exercise requiring high-power outputs.

Pre-Cooling and Intermittent-Sprint Performance
Previous pre-cooling literature has somewhat divergently highlighted the improvements observed in endurance performance, yet potentially negative effects on sprint or power-based activities. Given the required involvement of prolonged sub-maximal work interspaced with high-intensity efforts in intermittent-sprint exercise, it may be questioned what benefits pre-cooling could provide for team sports or intermittent-sprint exercise. However, intermittent-sprint activity is well documented to invoke larger thermoregulatory loads through higher core temperatures than continuous exercise [36] and accordingly may have an even greater need for pre-cooling interventions.

Based on previous research, the traditional view has been that there is little evidence to suggest that pre-cooling is beneficial for intermittent-sprint exercise [8, 36, 37]. Drust et al. [36] employed a 60-min cold shower to induce a 0.6°C difference in the pre-cooled state prior to 90-min soccer specific exercise protocol in ambient temperatures of 26°C. They reported no changes in total distance covered, heart rate, ventilation, oxygen consumption, RPE or blood lactate and glucose measures for a simulated soccer protocol. Duffield et al. [8] used a short-duration cooling intervention with an ice-vest before (5 min) and during (10 min) an 80-min intermittent-sprint protocol in 30°C heat. Given the limited extent and duration of the cooling exposure it is not surprising that pre-cooling had little effect on core or mean skin temperature, sweat loss, perceived exertion or blood lactate measures. Further, similarly to Drust et al. [36] there were no differences in work or power for intermittent-sprint activity due to pre-cooling. Cheung and Robinson [37] utilized 5°C coolant perfused through a full upper-body garment prior to 30 min of a 10-second cycle ergometer sprints separated by 5 min at 50% $VO_{2\,max}$. While the pre-cooling intervention reduced core temperature (0.5°C), minimal differences were evident between cooled and non-cooled conditions for mean peak power, heart rate, VO_2 and blood lactate values. While this collection of findings indicates little ergogenic benefit of pre-cooling, the respective use of protocols consisting of maximal interspersed with constant-intensity exercise may account for these findings. As evidence for this, Kay et al. [21] reported that pre-cooling affected the pacing strategy adopted when free-paced exercise was allowed.

Castle et al. [16] were the first to indicate potential benefits of pre-cooling for team-sport oriented, intermittent-sprint exercise reporting the maintenance of a 4% greater peak power output following cooling of the lower body with ice-packs. Cooling interventions included an upper-body ice-vest (10°C), water immersion (17°C) or ice-packs to the lower body (-16°C). Core and muscle temperatures were lower in both the water and ice-pack conditions, yet power and work were only elevated above the control condition with ice-pack cooling of the lower body. This is the first study to indicate the maintenance of a higher peak power when cooling the working musculature and is in contrast to previous theories of cooling [35, 37]. The authors postulated that given the lack of differences in VO_2 or markers of metabolism between conditions, the reduction in both muscle and core temperature resulted in a lower physical strain index. Further, Castle et al. [16] hypothesised that the optimal combination of a reduction in muscle and core temperatures and maintenance of sensory feedback of motor-unit activation may have resulted in improved performance in the ice-pack condition.

Duffield and Marino [24] used a protocol of repeated maximal 15-meter sprint efforts over an 80-min protocol in 32°C heat, however, rather than standardising the intensity of work between sprints, allowed free-paced activity. In doing so, this replicated the type and choice of activity that exists in game environments. As with previous research, repeated sprint ability was not improved following pre-cooling (ice-bath or ice-vest), however, the distance of sub-maximal work covered was greater following pre-cooling in an ice-bath (11°C for 15-min). They reported similarities between the results of improvements noted in previous endurance research [9, 29] and the improvements in distance covered during sub-maximal bouts interspacing the maximal efforts. These results, taken in the context of findings by Kay et al. [20], indicate that the mechanisms that result in the improvement in exercise in the heat following pre-cooling may be common, regardless of the type of exercise. Further, given the change in self-selected pacing strategies employed and lack of difference in cardiovascular or blood metabolite variables, the ergogenic benefits of pre-cooling may be more likely as a result of altered muscle recruitment patterns from a CNS origin.

In summary, the act of imposing an effective cooling intervention of sufficient extent and duration to lower skin, muscle and core temperature is beneficial for endurance and intermittent-sprint exercise performance, yet equivocal and likely to be negligible for power based performance [4, 24, 35]. Therefore, while the ergogenic benefits of pre-cooling for prolonged exercise seem to be evident and well documented, the actual mechanisms responsible for these improvements are somewhat equivocal and require further discussion.

Proposed Mechanisms of Pre-Cooling
Despite the extent of evidence on the ergogenic effects of pre-cooling, proposed mechanisms are few and traditionally relate to an overloading of cardiovascular and metabolic systems [4, 5]. As pre-cooling improves performance, with previous research highlighting an increase in O_2 delivery and reduction in heart rate, it has been presumed that these mechanisms resulted in the observed performance improvements [18, 21, 27], and continue to appear in recent pre-cooling literature [9]. However, the role of CNS control of exercise in response to [38] or in anticipation of [39] changes in the internal thermal load is now of greater focus. Accordingly, the mechanisms underpinning performance improvements following pre-cooling are also likely to originate from supraspinal dictation rather than cardiac or skeletal muscle function [20, 40].

Pre-cooling can increase the time to reach critical core temperatures, increase the rate of work and time to fatigue, increase the heat storage capacity, delay or reduce sweat loss and improve thermal comfort [10]. While implicit in these responses are reduced cardiovascular and metabolic strain, the differences in these factors in a pre-cooled, neutral or hyperthermic condition do not sufficiently explain performance decrements or improvements [38]. Further, tissue oxygenation and blood volumes have actually been reported to be reduced with cooling in non-exercising single limb models [17]. Accordingly, while peripheral mechanisms have some role in heat-induced fatigue, the key mechanisms of performance decrement are thought to be located in the CNS and result in a reduction in muscle activation or recruitment in response to a high internal thermal load [41]. These responses involve muscle specific reductions in sustained muscle activation [38], reduction in peak evoked torque and fibre recovery time [42], increased inhibitory sensory feedback from peripheral receptors [43], reduction in cerebral O_2 or glucose supply [44], reduction in cerebral neurochemical transmitters and altered balance of dopamine and serotonin [44] or an increased uptake of branched chain amino acids (BCAA) due to low glucose or O_2 supply [45]. Therefore, it is possible that pre-cooling may directly or indirectly influence one or all of these factors to consequently influence performance.

Pre-cooling improves the rate of heat storage and blunts the increase in thermal load which may reduce the extent of inhibitory feedback, assist maintenance of the equilibration of the blood-brain barrier control of O_2, glucose or BCAA concentrations or anticipate the slower rise in core temperature and delay the heat-induced down-regulation of muscle recruitment [39, 41, 47]. As such, in response or in anticipation of the improved heat storage, the CNS permits the selection of higher exercise intensities, especially during the time window of cooling. Hence, altered pacing strategies result in more work performed with concomitant increments in physiological perturbations, in particular core or brain temperature, that are similar to non-cooled responses [20, 24].

The suppression of core temperatures during exercise may allow the selection of higher self-selected work rates following pre-cooling [9, 20, 24]. While this response is not evident in constant-intensity exercise, the selection of higher work rates is present in free-paced continuous [20] and intermittent-sprint [24] exercise. As evidenced in figure 1, it should be noted that while sprint times do not differ, there are improved pacing strategies to complete more work in the ice bath condition. The similarities in selected pacing strategies in different types of exercise may indicate that the mechanisms underlying pre-cooling are linked to the selection of higher work rates due to a prevention of the reduction in muscle activation or recruitment, although there is limited EMG data to support this notion [47]. Central fatigue induced by hyperthermia may be thought of as a progressive inhibition of motor activation [41, 44], and accordingly the maintenance of a higher activation in a cooler state may be in response to the blunted core or brain temperatures. This point is further enforced by the observations of a finite duration of physiological responses to pre-cooling and similarity in core temperatures and cardiovascular and metabolic perturbations at exhaustion or end-exercise, despite a greater volume of work performed [9, 20, 24]. In summary, while research is lacking, it is proposed that the act of pre-cooling delays the inhibition of voluntary motor activation resulting from high or increasing core temperature, which allows the selection of higher exercise intensities and promotes more favourable pacing strategies to improve exercise performance in the heat.

Post-Cooling for Recovery from Exercise in the Heat

Exercise-Induced Thermal Load
The increased demands of competitive sport have witnessed the use of post-exercise cold water immersion and other cooling procedures to aid recovery for ensuing exercise bouts [11]. It is somewhat surprising that limited research attention has been paid to the recovery from exercise-induced heat stress, especially given the well-documented findings regarding higher physiological strain and reduced performances in hot conditions [14]. This may be somewhat accounted for by the predominance of laboratory-based data inducing high thermal stress during

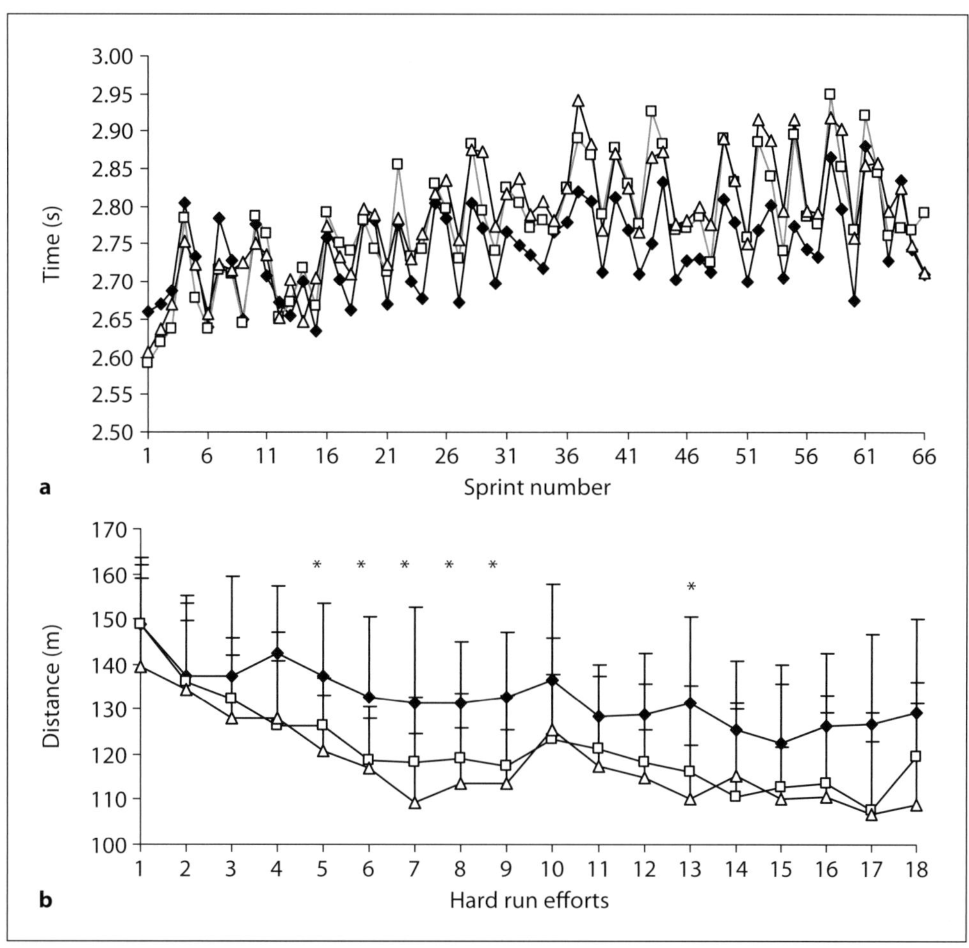

Fig. 1. a Mean 15-meter sprint time (s) for each condition and sprint number. **b** The hard running distance (m) for each condition for the hard run efforts. Conditions are ice-bath (■), ice-vest (□) and control (△) during intermittent-sprint exercise at 33°C. *Significantly different to ice-bath condition. **b** Data are redrawn from Duffield and Marino [24].

fixed-intensity exercise that may not be replicated in field environments. While constant-load exercise to exhaustion in a laboratory setting may induce high core temperatures, free-paced activity in the field is adjusted based on perceptual and physiological responses to ensure work is completed without excessive thermal strain [39, 40].

Recent improvements in technology have allowed greater insights to thermoregulatory responses in free-pace competitive environments [48, 49]. Laursen et al. [48] reported only moderate increments ($\sim$1°C) in core temperature following an ultra-endurance event in Ironman athletes in mild temperatures (17–26°C), without associations between core temperature, performance or dehydration. Further, previous team-sport field-based data has indicated a plateau in core temperature (below 39.3°C) during intermittent-sprint competition [49]. Additionally, unpublished observations by Duffield and Coutts highlighted this plateau in core temperature coincided with an overall reduction in distance covered (as measured by GPS). Moreover, there was a seemingly protective down-regulation in the intensity and volume of low to moderate-intensity work that ensured the maintenance of manageable core temperatures. Common findings of these studies indicate mean core temperatures were below 39.3°C, although several players can reach core temperatures of 40°C and accordingly, cooling may be an aggressive form of preventing any ensuing decline in performance and health [12]. Collectively, these data indicate that the relationship between work and internal thermal load is present in the field and that while the response to free-paced work is somewhat altered, high core temperatures can result that may affect current and ensuing bouts of exercise.

Suppression of Performance Post-Exercise in the Heat
High internal thermal loads may result in the decrement of acute exercise performance and delay the recovery back to optimal functioning [2]. Heat intolerance has been linked to general fatigue, sleeplessness and reduced performance [1]. Thus, the degree of exercise-induced heat stress or inability to tolerate imposed loads may result in prolonged suppression of exercise performance. Martin et al. [50] have highlighted the reduction in voluntary activation post-exhaustive exercise due to whole-body hyperthermia, implicating a selective CNS reduction of exercised muscles following exercise. In addition, Thomas et al. [51] have shown that passive heating resulted in the reduction of peak voluntary force due to increases in core temperature rather than specific muscle temperature. The return to optimal performance was speeded when cooling methods were introduced to increase the rate of reduction in core temperature. This may be a result of reduced CNS drive to active musculature as a preventative mechanism to reduce the production of metabolic heat and further act as a heat sink to blunt the rise in core temperature [40, 50, 51]. Moreover, it highlights that the maintenance of a high internal thermal load may affect performance in ensuing bouts of exercise.

Cooling Interventions Post-Exercise
Despite a lack of research on the influence of post-exercise cooling of recovery from exercise performance, the threat of delayed recovery and resultant effects on sleep, electrolyte balance and health are well documented [1, 14]. Previously Schniepp et al. [52] have reported that cold water immersion reduced both power output and heart rate when performed between two 30-second cycle ergometer sprints in moderate laboratory temperatures. This reduction in acute performance

fits with previous data indicating that cooling may have negative effects on maximal power [33]. In contrast, Lane and Wegner [53] reported a small increase in work performed during a second maximal 18-min cycling bout 24 h following an initial bout which was followed by cold water immersion. This is in contrast to the significant reductions noted when passive recovery was employed following the initial cycling bout. Vaile et al. [13] recently reported that active recovery resulted in a 4% reduction in ensuing performance, while cold water immersion maintained performance following recovery in the heat. Two respective 30-min cycling bouts were performed, separated by a 15-min recovery intervention (passive or a range of cold water temperatures) and 40 min passive seating in the heat (34°C). The cold water immersion conditions maintained exercise performance and reduced core and mean body temperatures and heart rate; however, there were no differences between conditions (active or any cold exposure) in heart rate or lactate values in the second exercise bouts. The authors postulated that due to all cold water temperatures (15–20°C) decreasing mean body temperature, a resultant adjustment in circulatory distribution occurred which may aid ensuing exercise performance.

While contradictory results exist, the act of post-exercise cooling is certainly beneficial for the reduction in the internal thermal load, and may be of benefit for endurance or prolonged high-intensity exercise performance in the short (1–2 h) or intermediate (24 h) recovery time frames. However, to date the mechanisms remain speculative and deserve some attention.

Proposed Mechanisms of Post-Cooling

As with any cooling procedure, commonly proposed mechanisms relate to an increase in circulatory volume and therefore an improved muscle blood flow [4, 13]. The resultant peripheral vasoconstriction following changes in skin temperature from cooling and closing of local vasculature promote a larger central circulatory reserve that may be beneficial to reduce cardiac stress due to the increased venous return and stroke volume [26]. While the cardiovascular explanation is certainly plausible, of the limited studies that have investigated post-exercise cooling, few studies report any differences in heart rates that are not explained by the passive nature of cold water immersion or differences in any cardiovascular measure during ensuing bouts of exercise [13, 52, 53]. It is also questionable that the increase in central blood volume resulting from cooling is sufficient to actually result in changes in cardiodynamic function [24, 38, 41]. Further, it seems somewhat counter-intuitive that to cool the periphery to increase the central blood volume allows a greater reserve to then re-direct back to the peripheral musculature to aid recovery of an area that is under local vasoconstriction. Finally, as previously mentioned tissue oxygenation and blood volumes have been reported to be reduced with cooling in non-exercising single limb models and remain that way for up to 60 min post-cooling [17].

A further explanation of the potential benefits noted following cooling for exercise relates to the prolonged suppression of CNS recruitment of voluntary force due to hyperthermic conditions [50] and result in earlier fatigue [14]. High core as opposed to muscle specific thermal loads result in a selective reduction of voluntary activation [50, 51]. The alleviation of this thermal stress by post-exercise cooling may assist in the return of voluntary force production and exercise performance. In agreement, Morrison et al. [46] have demonstrated that passive elevations in core temperature result in reduced voluntary activation that is reversed with cooling. Given the absence of notable differences in cardiovascular, metabolite or hormonal measures due to cooling interventions in the recent research [13], it is possible that centrally mediated mechanisms based on the internal thermal load regulate the recovery of exercise performance following exercise-induced heat stress. However, to date this is speculative as there is currently no research to support or refute this notion.

Finally, the reduction of core temperature may also have added benefits that promote faster recovery from the heat over longer recovery durations. Optimal core temperatures exist that promote the onset and quality of sleep, and accordingly reductions in core temperature also assist in the improvement in sleep quality, while also improving perceptions of exercise and resting fatigue [1, 3]. Post-exercise cooling may also reduce the sensations of pain [12] and inflammatory response to any induced muscle damage or injury [11]. While these factors have not been directly attributed to improved exercise performance, it is likely that singularly or as an additive effect they act to assist both the acute or prolonged recovery of performance following exercise-induced heat stress.

Conclusion

In summary, the research evidence from both laboratory and field environments highlight the reduction of performance as a result of exercise-induced heat stress. A current and growing body of literature indicates that the use of cooling methods can improve both performance of, and recovery from, any prolonged low or high-intensity exercise. While popular notions of increased central blood volume and muscle blood flow remain entrenched in the research literature, the evidence and theory behind such explanations is not as strong as previously thought. While alteration to local and systemic blood flow are highly likely, they may actually be a result of centrally controlled regulation in response to muscle recruitment. Growing evidence indicates a complex role of central regulation of voluntary muscle activation, and therefore performance, in response or anticipation of the internal thermal load. Accordingly, the act of pre- or post-exercise cooling may alleviate this thermal load and allow an increase in the activation and recruitment of muscle force, thereby improving the performance and recovery. Regardless of the mechanisms, there is sufficient evidence to regard pre- and post-exercise cooling as an important ergogenic aid and recommend the use of such procedures in the protection and recovery of exercise-induced heat stress.

References

1 Armstrong LE, Casa DJ, Millard-Stafford M, Moran DS, Pyne SW, Roberts WO: American College of Sports Medicine position stand: exertional heat illness during training and competition. Med Sci Sports Exerc 2007;39;556–572.

2 Wendt D, van Loon LJC, van Marken Lichtenbelt WD: Thermoregulation during exercise in the heat. Sports Med 2007;37:669–682.

3 Maughan R, Shirreffs S: Exercise in the heat: challenges and opportunities. J Sports Sci 2004;22: 917–927.

4 Marino FE: Methods, advantages and limitations of body cooling for exercise performance. Br J Sports Med 2002;36:89–94.

5 Quod MJ, Martin DT, Laursen PB: Cooling athletes before competition in the heat: comparison of techniques and practical considerations. Sports Med 2006;36:371–382.

6 Hessemer V, Langusch D, Bruck K: Effect of slightly lowered temperatures on endurance performance in humans. J Appl Physiol 1984;57:1731–1737.

7 Wilson TE, Johnson SC, Petajan JH, Davis SL, Gappmaier E, Luetkemeier MJ, White AT: Thermal regulatory responses to submaximal cycling following lower-body cooling in humans. Eur J Appl Physiol 2002;88:67–75.

8 Duffield R, Dawson B, Bishop D, Fitzsimons M, Lawrence S: Effect of wearing and ice cooling jacket on repeat sprint performance in warm/humid conditions. Br J Sports Med 2003;37:164–169.

9 Arngrímsson SA, Petitt DS, Stueck MG, Jorgensen DK, Cureton KJ: Cooling vest worn during active warm-up improves 5-km run performance in the heat. J Appl Physiol 2004;96:1876–1874.

10 Reilly T, Drust B, Gregson W: Thermoregulation in elite athletes. Curr Opin Clin Nutr Metab Care 2006; 9:666–671.

11 Barnett A: Using recovery modalities between training sessions in elite athletes: does it help? Sports Med 2006;36:781–796.

12 Meeusen R, Lievens P: The use of cryotherapy in sports injuries. Sports Med 1986;3:398–414.

13 Vaile J, Halson S, Gill N, Dawson B: Effect of cold water immersion on repeat cycling performance and thermoregulation in the heat. J Sports Sci 2008; 26:431–440.

14 Gonzalez-Alonso J, Teller C, Anderson SL, Jensen FB, Hyldig T, Nielsen B: Influence of body temperature on the development of fatigue during prolonged exercise in the heat. J Appl Physiol 1999;86: 1032–1038.

15 Drust B, Rasmussen P, Mohr M, Nielsen B, Nybo L: Elevations in core and muscle temperature impairs repeated sprint performance. Acta Physiol Scand 2005;183:181–190.

16 Castle PC, Macdonald AL, Philip A, Webborn A, Watt PW, Maxwell NS: Precooling leg muscle improves intermittent sprint exercise performance in hot, humid conditions. J Appl Physiol 2006;100: 1377–1384.

17 Yanagisawa O, Homma T, Okuwaki T, Shimao D, Takahashi H: Effects of cooling on human skin and skeletal muscle. Eur J Appl Physiol 2007;100: 737–745.

18 Lee D, Haymes E: Exercise duration and thermoregulatory responses after whole body precooling. J Appl Physiol 1995;79:1971–1976.

19 Daanen HA, van Es MA, de Graaf JL: Heat strain and gross efficiency during endurance exercise after lower, upper, or whole body precooling in the heat. Int J Sports Med 2005;27:379–388.

20 Kay D, Taafe DR, Marino FE: Whole-body precooling and heat storage during self paced cycling performance in warm humid conditions. J Sports Sci 1999;17:937–944.

21 Olschewski H, Bruck K: Thermoregulatory, cardiovascular, and muscular factors related to exercise after precooling. J Appl Physiol 1988;64:803–811.

22 Hasegawa H, Takatori T, Komura T, Yamsaki M: Wearing a cooling jacket during exercise reduces thermal strain and improves endurance performance in a warm environment. J Strength Cond Res 2005;19:122–128.

23 Clements JM, Casa DJ, Knight JC, McClung JM, Blake AS, Meenen PM, Gilmer AM, Caldwell KA: Ice-water immersion and cold-water immersion provide similar cooling rates in runners with exercise-induced hyperthermia. J Athletic Training 2002; 37:146–150.

24 Duffield R, Marino FE: Effects of pre-cooling procedures on intermittent-sprint exercise performance in warm conditions. Eur J Appl Physiol 2007; 100:727–735.

25 Grahn DA, Cao VH, Heller HC: Heat extraction through the palm of one hand improves aerobic exercise endurance in a hot environment. J Appl Physiol 2005;99:972–978.

26 Maw GJ, Mackenzie IL, Taylor NAS: Can skin temperature manipulation, with minimal core temperature change, influence plasma volume in resting humans. Eur J Appl Physiol 2000;81:159–162.

27 Schmidt V, Brück K: Effects of a precooling manoeuvre on body temperature and exercise performance. J Appl Physiol 1981;50:772–778.

28 Mitchell JB, McFarlin BK, Dugas JP: The effect of pre-exercise cooling on high intensity running performance in the heat. Int J Sports Med 2003;24: 118–124.

29 Booth J, Marino F, Ward JJ: Improved running performance in hot humid conditions following whole body precooling. Med Sci Sports Exerc 1997; 29:943–949.

30 Cotter JD, Sleivert GG, Roberts WS, Febbraio MA: Effect of precooling, with and without thigh cooling, on strain and endurance exercise performance in the heat. Comp Biochem Physiol [A] 2001;128: 667–677.

31 Tucker R, Rauch L, Harley YX, Noakes TD: Impaired exercise performance in the heat is associated with an anticipatory reduction in skeletal muscle activation. Pflügers Arch 2004;448: 422–430.

32 Marsh D, Sleivert G: Effect of precooling on high intensity cycling performance. Br J Sports Med 1993;33:393–397.

33 Holewijn M, Heus R: Effects of temperature on electromyogram and muscle function. Eur J Appl Physiol Occup Physiol 1992;65:541–545.

34 Crowley GC, Garg A, Lohn MS, Van Someren S, Wade AJ: Effects of cooling the legs on performance in a standard Wingate anaerobic power test. Br J Sports Med 1991;25:200–203.

35 Sleivert GG, Cotter JD, Roberts WS, Febbraio MA: The influence of whole-body vs. torso precooling on physiological strain and performance of high intensity exercise in the heat. Comp Biochem Physiol 2001;128:657–666.

36 Drust B, Cable NT, Reilly T: Investigation of the effects of the pre-cooling on the physiological responses to soccer-specific intermittent exercise. Eur J Appl Physiol 2000;81:11–17.

37 Cheung SS, Robinson AM: The influence of upper-body pre-cooling on repeated sprint performance in moderate ambient temperatures. J Sports Sci 2004;22:605–612.

38 Nybo L, Nielsen B: Hyperthermia and central fatigue during prolonged exercise in humans. J Appl Physiol 2001;91:1055–1066.

39 Marino FE: Anticipatory regulation and avoidance of catastrophe during exercise-induced hyperthermia. Comp Biochem Physiol Part B 2004;139: 561–569.

40 Tucker R, Marle T, Lambert EV, Noakes TD: The rate of heat storage mediates an anticipatory reduction in exercise intensity during cycling at a fixed rating of perceived exertion. J Physiol (Lond) 2006; 574:905–915.

41 Nybo L: Exercise and heat stress: cerebral challenges and consequences. Prog Brain Res 2007;162: 29–43.

42 Todd G, Butler JE, Taylor JL, Gandevia SC: Hyperthermia: a failure of the motor cortex and the muscle. J Physiol 2005;563:621–631.

43 Amann M, Eldridge MW, Lovering AT, Stickland MK, Pegelow DF, Dempsey JA: Arterial oxygenation influences central motor output and exercise performance via effects on peripheral locomotor muscle fatigue in humans. J Physiol (Lond) 2006; 575:937–952.

44 Nybo L: Hyperthermia and fatigue. J Appl Physiol 2008;104:871–878.

45 Nybo L, Secher NH: Central perturbations provoked by prolonged exercise. Prog Neurobiol 2004; 72:223–261.

46 Morrison S, Sleivert GC, Cheung SS: Passive hyperthermia reduces voluntary activation and isometric force production. Eur J Appl Physiol 2004; 91:729–736.

47 Hunter AM, St Clair Gibson A, Mbambo Z, Lambert MI, Noakes TD: The effects of heat stress on neuromuscular activity during endurance exercise. Pflügers Arch 2002;444:738–743.

48 Laursen PB, Suriano R, Quod MJ, Lee H, Abbiss CR, Nosaka K, Martin DT, Bishop D: Core temperature and hydration status during an Ironman triathlon. Br J Sports Med 2006;40:320–325.

49 Edwards A, Clark NA: Thermoregulatory observations in soccer match play: professional and recreational level applications using an intestinal pill system to measure core temperature. Br J Sports Med 2006;40:133–138.

Duffield

50 Martin PG, Marino FE, Rattey J, Kay D, Cannon J: Reduced voluntary activation of human skeletal muscle during shortening and lengthening contractions in whole body hyperthermia. Exp Physiol 2004;90:225–236.

51 Thomas MM, Cheung SS, Elder GC, Sleivert GG: Voluntary muscle activation is impaired by core temperature rather than local muscle temperature. J Appl Physiol 2006;100:1361–1369.

52 Schniepp J, Campbell TS, Powell KL, Pincivero DM: The effects of cold-water immersion on power output and heart rate in elite cyclists. J Strength Cond Res 2002;16:561–566.

53 Lane KN, Wegner HA: Effect of selected recovery conditions on performance of repeated bouts of intermittent cycling separated by 24 hours. J Strength Cond Res 2004;18:855–860.

Rob Duffield, PhD
School of Human Movement Studies and Exercise and Sports Science Laboratories
Charles Sturt University
Panorama Ave, Bathurst, NSW 2795 (Australia)
Tel. +61 2 6338 4939, Fax +61 2 6338 4065, E-Mail rduffield@csu.edu.au

Marino FE (ed): Thermoregulation and Human Performance. Physiological and Biological Aspects.
Med Sport Sci. Basel, Karger, 2008, vol 53, pp 104–120

Ethnicity and Temperature Regulation

Michael I. Lambert[a] · Theresa Mann[a] ·
Jonathan P. Dugas[b]

[a]MRC/UCT Research Unit of Exercise Science and Sports Medicine,
Department of Human Biology, University of Cape Town, Newlands,
South Africa; [b]Department of Kinesiology and Nutrition,
University of Illinois at Chicago, Chicago, Ill., USA

Abstract

There are at least 31 climatic zones around the world ranging from year-round freezing conditions to daily hot temperatures of around 45°C. Each zone is inhabited by people who have adapted their lifestyles to accommodate the environmental conditions. There are many examples showing physiological and morphological differences between groups living in different environmental conditions (i.e. climate has been shown to influence characteristics including birth weight, body shape and composition, cranial morphology and skin color and sensitivity). Whilst the phenotypic differences are very clear, the genotypic differences are less easy to discern. This can be attributed to the logistical difficulties in executing the definitive study which controls for the environmental and lifestyle factors which themselves induce physiological and morphological changes. However, considering that at least 50 genes have been identified which have altered expression after exposure to heat and at least 20 genes are altered by cold exposure, it is reasonable to assume that more physiological and morphological differences will be attributed to genetic origins as the data becomes available.

Copyright © 2008 S. Karger AG, Basel

The diversity of climates around the world, and their impact on populations living in them for long and short periods has interested scientists for centuries. The first modern classification of the various world climates was published by Wladimir Köppen in 1900 [1]. He classified the planet into five main zones (equatorial, arid, warm temperature, snow, and polar). After including subsections for differences in precipitation and temperature, this classification identifies 31 different climate zones with a gradual transition in conditions from zone to zone. An example of extreme conditions is the Arctic, where winters are long and cold with only brief, cool summers. Ice covers the ocean most of the year causing subfreezing temperatures. The average temperature is about $-7°C$ with minimum temperatures reaching about $-68°C$. These cold conditions can be contrasted to the hot temperatures of the deserts which reach highs of around $45°C$.

Humans *(Homo sapiens)* are the only species that have adapted to live and function in these very different environmental conditions. Although questions about the interaction between climate and ethnic diversity remain unanswered, scientists probing back up to 40,000 years have offered several likely explanations [2]. Indeed, even some of the earliest research begins to shed light on the human ability to adapt to extreme environmental conditions.

Paleoclimatology research (i.e. the study of the climate of past ages) has shown that the average temperature in the Netherlands 50,000 to 41,000 years ago was estimated to be $-1°C$ [2]. It was during this time that Neanderthals roamed Europe and the general belief was that humans only colonized the region in the last stages of the Pleistocene Ice Age, some 13,000–14,000 years ago [3]. More contemporary views are that Europe was inhabited by humans long before the Neanderthal vanished from the continent about 28,000 years ago [3, 4]. While it is beyond the scope of this chapter to debate the finer details about the origin of humans [5], it can be assumed as fact that wide climatic variations occurred during the ice age. The ability of humans to adapt in the face of extreme environmental changes was a factor which contributed to the survival of the species [4].

Adaptation to the changing environment was a consequence of physiological plasticity. Long-term adaptations resulted in people changing their size and shape in response to the environment [6]. This phenomenon was first observed by Carl Bergmann, a German biologist who reported that within the same species of warm-blooded animals, populations with smaller individuals are usually found in warm climates near the equator while larger individuals within a population are found further from the equator in colder regions [7]. Thirty years later, Joel Allen noted within the same species that individuals in populations living in warm climates had longer limbs than those living in colder climates [8]. Does this apply to humans?

Studies have shown great diversity between groups living in different regions. For example, the average height of Pygmies of Africa was 144.9 cm (men) and 136.1 cm (women) [9] compared to Dutch men (184.0 cm) and women (170.6 cm) [10]. Are these differences due to climatic adaptations or ethnic differences arising from other factors? Clues of the impact of the environment on subsequent morphological development can be obtained from a study of a large sample (n = 97,237) including 140 populations in which it was shown that there was an inverse relationship between the annual heat index of the region in which a population lived and the mean birth weight of the population (r = -0.59) [11]. This study shows that the interaction between environment and growth may begin in the uterus where the hot environment retards the growth and development of the foetus [11].

Before we explore this question of the interaction between climate and ethnicity further, we need to define 'ethnicity' – the term has been misused in the past when it has been inferred that ethnic identities are associated with genetic differences [12]. To avoid ambiguity, ethnicity in this chapter is defined as the social formation based on shared culture, values, language, descent, religion or other commonality [13].

The aim of this chapter is to discuss the interaction between thermoregulation and ethnicity. In particular, we have attempted to explain the short-term physiological adaptations which occur after exposure to either hot or cold conditions, followed by a discussion on physiological adaptations which occur after long-term exposure to extreme climates. Whenever possible we will draw attention to possible genotypic differences.

Acute Adaptations

Physiological adaptations as a consequence of exercise training have been widely studied, resulting in a body of knowledge about the nature of the adaptation, the time course and the decay of these adaptations once the 'stress' of training is removed. The adaptations to both cold and especially hot environments are equally well researched. The important points arising from this work are presented.

Acclimation to a Hot Environment

The earliest scientists to examine heat acclimation were those such as Buskirk [14] and Adolph [15], but many other investigators have contributed to this area. The first important finding of their collective work was that humans have the capacity to adapt after living and working in a hot environment. This initial insight then lead to more detailed investigations which have provided a good understanding of the detail and nature of the adaptations.

The adaptations are numerous and varied, but the net result is that exercise performance in the heat improves following a period of acclimation. This improvement can be marked, and endurance time to fatigue can nearly double [16]. Several different mechanisms are responsible for this improvement in thermoregulatory function, but the net effect of these changes is that the individual gains a better ability to cope with the physiological stressors associated with a hot environment.

Circulatory Changes

As is the case with exercise training, a plasma volume expansion is one of the first adaptations which occurs after exposure to a hot environment [17–20], and after a heat acclimation protocol it would not be unusual for increases in plasma volume of greater than 10% [16]. This adaptation leads to other circulatory changes, including heart rates that are 5–10 bpm lower following acclimation when exercising in the heat [16, 21]. Accompanying the changes in plasma volume is also an enhanced cardiac output and generally enhanced function of the cardiovascular system to move blood around the body. Therefore, following acclimation one can expect to see increases in cardiac output in the range of 3–5%. This is largely the result of a significant increase in stroke volume of approximately 17% [16].

One demand which the body must meet during exercise in a hot environment is an increased dissipation of metabolic heat. This requires an increase in skin blood flow, with forearm blood flow increasing by as much as 15% following acclimation [16]. Therefore, more blood is sent to the surface tissues so that the heat can be transferred to the skin and sweat for radiation and evaporation. However, in spite of this increased demand for blood flow to the skin, it should be noted that the other changes that occur with acclimation help to reduce any undue stress from being placed on the cardiovascular system. These adaptations allow the person to cope better with the hot environmental conditions.

Sweating Responses

The circulatory changes mentioned above all contribute to aiding the body in its challenge to dissipate its metabolic heat in a hot environment. An additional adaptation that humans make during acclimation is that sweating response changes significantly. The sweating response is essential for thermoregulation and the maintenance of body core temperature as evaporation of sweat from the skin is the dominant means of heat loss during exercise. In a pre-acclimatized state the sweating response begins much later than after acclimation, suggesting that exposure to a hot environment increases the sensitivity of this response. When fully acclimated, the sweating response is modulated in two significant ways. First, sweating will begin much sooner following the onset of exercise. Second, sweat production is enhanced so that a larger volume of sweat (up to 15%) is produced [16].

Maintenance of Body Core Temperature

The primary consequence of the adaptations outlined above is that the body core temperature is maintained within a desirable range. Due to the adaptations there is a largely enhanced ability to dissipate heat and regulate temperature, and a positive outcome of this is that an acclimatized person can tolerate a greater heat load and perform more work in the heat.

Prior to acclimation, the core temperature rises more steeply and reaches an individually critical temperature, at which time fatigue occurs. However, following acclimation, the rise in core temperature is less steep and therefore the individual can exercise longer before reaching an individually limiting core temperature [21].

Time Course of Adaptations and Decay

The time course and magnitude of the adaptations will be a function of both the frequency and the intensity of the exposure to the heat conditions. However, a typical heat acclimation protocol consists of daily exercise for 60–90 min at $\geq$50% $VO_{2\,max}$. When exposed to this program, the cardiovascular adaptations are achieved relatively quickly and are complete within 1–5 days, while those adaptations that produce changes in sweating rate and ultimately core temperature require a full fourteen days to manifest [22]. Conversely, the decay of these adaptations can occur at a quicker rate and diminish in only a few days [22]. The loss of the adaptations can be attenuated with frequent exposures to a hot environment [22].

Acclimation to a Cold Environment

Cold acclimation has been well studied with much evidence supporting the conclusion that humans do make physiological adaptations to both cold water and cold air exposure [23–26]. Converse to the challenges faced when exposed to a hot environment, the primary challenge of a cold environment is for the body to conserve heat. This is accomplished both by circulatory changes to reduce heat losses and also via metabolic changes that increase heat production.

The time course of the adaptations to cold exposure is approximately 10 days, and consists first of cardiovascular changes that help to reduce heat losses to the environment. These changes involve constricting the peripheral and superficial vascular beds [26] so that the central blood volume and body heat is maintained around the vital organs. As the cold exposure is sustained over a number of days, it appears that humans can take advantage of metabolic adaptations to then increase heat production. These include higher circulating concentrations of noradrenalin and thyroxin, and an increase in circulating free-fatty acids, the result of which is enhanced non-shivering thermogenesis which results in an increase in heat production [26].

Summary of Acute Adaptations to Hot and Cold Conditions

Humans can adapt to both hot and cold environments, and the adaptations are similar in that cardiovascular changes appear first and rapidly. Chronic exposure leads to more robust neuro-humoral adaptations that further aid in defending the core temperature. The time course of the adaptations to both hot and cold conditions is relatively short, and the adaptations are complete within 2–4 weeks in most individuals. However, as occurs with the adaptations following exercise training, the thermoregulatory adaptations to hot and cold environments are lost quickly once the exposure ceases.

With this discussion on the acute adaptations to hot and cold conditions as background, the next section will discuss the physiological characteristics/adaptations that occur in different groups that have been exposed to these extreme environmental conditions for prolonged periods, in some cases several generations. The reader is also referred to a comprehensive review which discusses some aspects which are beyond the scope of this chapter [27].

Physiological Characteristics in Ethnic Groups that Reduce Thermal Strain

Body Mass

As mentioned in the introduction, Bergmann [7] and Allen [8] were the first scientists to describe the relationship between body 'size' and environmental temperature. Their findings became known as 'Bergmann and Allen's rules' and were originally applied to birds and mammals. Bergmann's use of the term 'body size' was somewhat ambiguous, so to clarify this concept a study was done just over a hundred years later to examine the relationship between body mass and environmental temperature. This study showed that the mean body mass in 116 male indigenous groups was inversely related to mean environmental temperature [28]. The lowest body mass was observed amongst the groups that had inhabited tropical areas for a considerable period, e.g. South Asian, Australian Aboriginal and African populations. Conversely, those people who were indigenous to America, Europe and Central Asia, countries with lower mean annual temperatures, had a higher body mass. Seventy percent of total variance in body mass between groups was ascribed to stature and temperature. The remaining variance depended on socio-economic and nutritional standards [28].

As was the case with geographically distant populations, it was observed that the body mass of racially and culturally 'similar' groups, such as the Ituri Bambuti, Ruanda Batwa and Bahutu, varied with changes in altitude and, by implication, temperature [28]. The variation in body mass between closely related groups of peoples in response to living in different temperatures would support the notion that this is primarily a phenotypic response. However, there are also several studies which suggest that there are also genetic influences as the original inhabitants of tropical areas exhibited a lower mean body weight than other groups which arrived later. For example, the aboriginal Orang Balik Papan of Borneo had a mean weight of 46 kg compared to the 51 kg of later immigrants [28].

Body Shape and Body Surface Area

Body surface area largely determines the capacity for heat loss. In humans a lower body surface area: body mass ratio is conducive to heat conservation. Each characteristic gains relative importance according to the environmental conditions. It follows that the ratio of surface area and body mass has significant implications for thermoregulation.

When comparing low and high body surface area:body mass ratios, it is typical to contrast the tall, linear form of an African (e.g. the Dinkas of East Africa) with the shorter, stockier build of an arctic Eskimo. Whilst this body type comparison is certainly valid, it is not the only one available [29]. Pygmies are, proportionately, even more thickset than Eskimos, yet their small stature enables them to retain a high surface area:body mass ratio. Moreover, it is appropriate to use Ruff's analogy of a body represented by a cylinder of a certain width. According to this model, while the width remains fixed, height may vary and the surface area:mass ratio remains the same [29]. Conversely, increasing the width of the cylinder will always reduce the ratio, independent of height. This model corresponds well with modern human populations living in different regions. Firstly, populations living in the same climatic zone have a similar body breadth. Consider the Nuer males, who are the tallest living Africans, and the Eastern Pygmy women, the smallest stature Africans. The pelvic width of both these groups are similar (within 3 cm) despite a difference in stature of about 48 cm. Moreover, those people living in cold climates generally have a noticeably wider body breadth compared to their tropical counterparts.

These findings suggest that breadth rather than stature is the more important factor for thermoregulation. However, stature does retain some significance. Specifically, a tall, linear body is advantageous in a dry, open environment as heat gain from the sun may be minimised whilst increased exposure to convective currents maximises sweat evaporation. Conversely, in a dim, humid environment such as the rain forests inhabited by the African Pygmies, sweat evaporation is restricted and the reduced heat production associated with a small body mass would be of greater benefit.

The Pacific Paradox

According to Bergmann and Allen's rules, one would expect the islanders of tropical Polynesia to display the linear body form and high surface area:body mass ratio typical of warm, low latitude regions. However, these individuals are amongst the largest and most muscular people on Earth. Philip Houghton, who has studied these populations extensively, attributes their apparent 'cold climate' morphology to the trade winds which blow steadily in the region from April to September. Often accompanied by overcast conditions, these winds create a cold, wet environment to which Polynesians would have been particularly exposed through their habitual sea voyages and coastal fishing [30]. In these uniquely cold surroundings, only those individuals with large, muscular body types would have been able to retain adequate body heat and be able to function over a prolonged period whereas leaner individuals would have suffered severely, perhaps fatally, from hypothermia [30]. Polynesians are believed to be descended from South East Asians who migrated to these oceanic islands by way of Indonesia and Melanesia

[31]. For this reason, Houghton believes that the muscular body type of the Polynesians only evolved after the islands were inhabited [30].

Skeletal Adaptations
Skeletal formation is clearly relevant for defining stature and body shape, and consequently an individual's body surface: body mass ratio, the importance of which has been discussed. Therefore it stands to reason that the finer details of skeletal morphology may also reveal significant information about the ability to thermoregulate. Cranial morphology, for example, has been associated with thermoregulation and it was proposed that a rounded head (brachicephaly) would be better suited to heat retention than an elongated head (dolichocephaly). However, this theory is undermined by the distinctly dolichocephalic skulls of Eskimo and Fueguian populations who tolerated the extremely cold conditions associated with latitudes of 70°N and 53°S, respectively [32]. Rather, it would appear that overall cranial capacity is a more valid comparison as this parameter has been shown to vary amongst populations [33, 34]. Indeed, the higher mean cranial capacity reported in Eskimos compared to Australian Aborigines supports the general perception that the limitation of the size of the brain is determined by the size that can be cooled [35]. As brain volume has no bearing on intellectual capacity within a species [34], a smaller cranium in a hot environment is a meaningful adaptation designed to avert the convulsions, fainting or even permanent damage triggered by even slight overheating of the brain.

In addition to cranium size, craniofacial characteristics may also hold significance for thermoregulation. Specifically, differences in nasal cavity size have been attributed to environmental influence [35]. For example, individuals from a hot climate tend to have a broad nasal cavity with maximum surface area for cooling inhaled air. Conversely, nasal bones of those people living in very cold climates, such as Eskimos and Fueguians, typically have a 'pinched up' appearance and an expanded interior nasal chamber [32]. A high, narrow nose warms and moistens inhaled air and, in addition, recovers heat and moisture from expired air [36]. Fueguians and Eskimos are morphologically distinct from all other present-day American Indian groups, largely as a result of differences in nasal height [32]. This distinction provides reasonable evidence of adaptation to a cold environment, although the role of genetics and differences in masticatory stress in determining craniofacial morphology cannot be entirely discounted [32].

Soft Tissue – Lips and Hair
It is interesting to note that lip size and hair form are also designed to facilitate heat balance in a particular environment. Thick, everted lips, typically observed in people of African origin, increase the surface area for heat loss when in a hot environment, whereas those individuals with thin, inverted lips will conserve body heat

better and reduce the likelihood of their lips freezing when conditions are particularly cold [37]. Furthermore, the tightly-coiled, wool-like hair observed, for example, in the Kalahari Bushmen, tends to aid heat loss from the scalp, keeping the brain cool under sweltering conditions. In cold climates, however, straight hair contributes to reducing heat loss from the scalp and keeping the brain sufficiently warm.

Skin Colour

Most would agree that skin colour is the most noticeable way in which races, and in some cases ethnicities, may be distinguished. Melanin pigmentation, the amount and type of which determines skin colour, affords dual photoprotection of dermal structures through its role as (1) an optical filter, scattering and thus attenuating absorption of radiation in the epidermis, and (2) a chemical filter – as a stable free radical, melanin can absorb potentially toxic or carcinogenic compounds produced by photochemical action [38].

The sweat glands in the dermis are protected from ultraviolet light (UV) radiation through the action of melanin. Sunburnt skin does not sweat effectively [39, 40] which can seriously compromise the body's most proficient mode of heat loss – evaporation. Furthermore, other deleterious effects of excessive UV radiation in the short-term include photolysis of folate, an essential nutrient involved in the biosynthesis and/or development of DNA, sperm, bone marrow, red blood cells and the embryonic neural tube [41–43]. Long term, degenerative changes in the dermis and epidermis may eventually develop into skin cancer. Clearly, a dark epidermis is of great advantage and not only for protecting sweat glands from UV-induced damage. However, exposure to UV radiation does confer one benefit in that it induces the synthesis of pre-vitamin D_3, essential for normal growth, calcium absorption and skeletal development. Hence, it is widely believed that populations which migrated outside of the tropics (e.g. Northern Scandinavians and Southern Africans [33]) evolved varying degrees of depigmentation as an adaptive mechanism to maintain adequate vitamin D_3 biosynthesis [44, 45]. Interestingly, the comparatively lighter skin tone ubiquitously observed in the females of a particular population is thought to facilitate the increased requirement for vitamin D_3 associated with pregnancy [46]. Jablonski and Chaplin [46], using remotely sensed data on UV radiation levels, were able to examine the geographic distribution of the potential for previtamin D_3 synthesis as well as the relationship between UV radiation levels and skin colour reflectances in indigenous populations. As anticipated, skin reflectance was strongly correlated with latitude and UV radiation levels [46]. This is shown in figure 1. Interestingly, differences in skin colour between the hemispheres have been attributed to the fact that the proportion of habitable land in the southern hemisphere that lies close to the equator is much larger than that in the northern hemisphere [47].

Secondly, variations in skin colour may evolve over time to regulate the penetration of UV radiation into the epidermis [46]. Indeed, all humans have a comparable

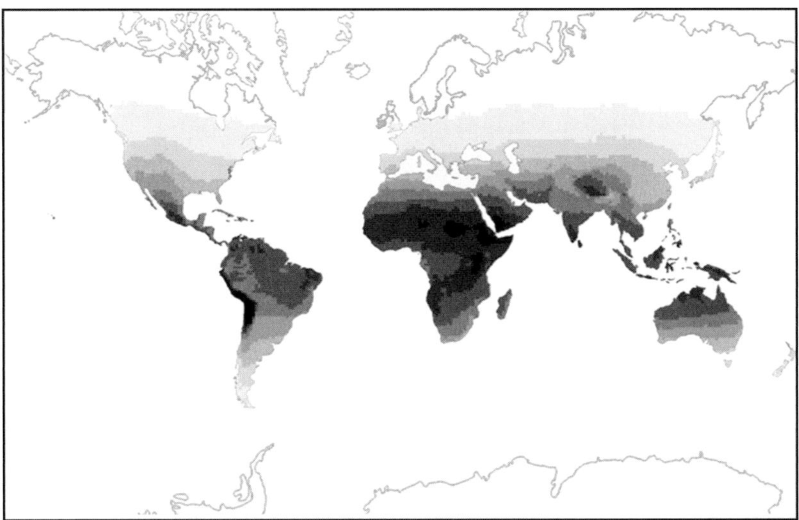

Fig. 1. Predicted skin colour of indigenous populations based on regression model of skin reflectance (at 685 nm). The darkest grey represents the populations with the greatest melanisation and the lightest grey represents the least melanisation. From Jablonski and Chaplin [46], with permission.

spatial density of melanin-producing melanocytes, it is only the activity of these melanocytes which differs from one racial group to another [37]. In this way, the degree of pigmentation observed in the indigenous inhabitants of a particular 'UV radiation regime' should optimally compromise between photoprotection and vitamin D synthesis [46]. In summary, traditionally dark-skinned ethnicities have an inherent advantage in maintaining somatic thermoregulation in the heat due to superior UV protection of their sweat glands. However, because these individuals require 2 to 6 times as much radiation as paler individuals to synthesize an equivalent amount of previtamin D [48, 49], they are at high risk of vitamin D deficiency should a change in lifestyle or locale reduce their exposure.

Skin Sensitivity to the Cold
The Eskimos and the Lapps are amongst several groups of people able to inhabit very cold environments due to technological intervention, notably thick, protective clothing. Indeed, the microclimate inside the clothing of the Lapps, or under the cloak of a Kalahari Bushmen sleeping by a campfire, has been found to be warm, or at least thermoneutral [50, 51]. Nevertheless, humans have previously been known to tolerate freezing conditions with minimal protection against the elements and it is widely believed that selective forces within these populations favoured extraordinary cold tolerance and the efficient use of body heat [52].

Perhaps the most extreme environment ever inhabited in this manner is Tierra del Fuego at the tip of South America, once home to four, now extinct, Fueguian tribes. Wearing only small capes made from mammal skins, Fueguians endured cold, rain, fog and almost constant wind with temperatures dropping as low as $-12°C$ in winter [32]. It stands to reason that these individuals must have developed several physiological adaptations to survive under such conditions and it is possible that reduced skin sensation, as perceived in high latitude, prehistoric Polynesians [52], may be such an example. Similar to the Fueguians, prehistoric Polynesians lived in a cool and wet marine environment, although a mean annual temperature of $11.1°C$ (around the southern latitudes of New Zealand) made the conditions marginally less severe. The large, muscular physique and short distal limb segments of the New Zealand Maori and the Chatham Islands Moriori would have assisted these individuals in body heat retention [27]. There was also speculation that neural adaptations might also have contributed to this remarkable tolerance of the cold [52]. As a consequence, five cranial nerve foramina through which nerves carrying cutaneous and deep sensory information entering the skull were studied. These features were of interest because the cross-sectional area of a nerve is directly proportional to the number of axons and hence the number of sensory receptors represented. It follows that fewer receptors would render an individual less sensitive to the cold and small changes in temperature [52]. After comparing the skulls of prehistoric Maori, prehistoric Moriori and contemporary Indian populations, it was reported that the infraorbital foramen was significantly smaller in the Maori and Moriori vs. Indians [52]. The nerves passing through this aperture carry facial cutaneous and sensory information and the postulated reduced sensory input in the cold-climate populations would have delayed the activation of heat producing/conserving mechanisms [52]. This increased 'cold threshold' has been advocated as a selective advantage in that by allowing the peripheral body temperature to drop, the energy required for thermogenesis in the limbs is reduced. When subsisting in a harsh environment where food was at a premium, this adaptation would have conserved valuable energy stores and allowed the individuals to be less uncomfortable. Interestingly, no difference was observed in the sensory nerve supply to the deep structures [52], leading to the conclusion that core temperature is strictly monitored irrespective of the prevailing climatic conditions.

Skin Sensitivity to Heat

In the same way that those people indigenous to cold climates may have increased their 'cold threshold', it may be argued that people inhabiting tropical regions have a reduced response to the heat. Specifically, there are several reports of a delayed onset of sweating and lower sweat volumes in tropical natives [53–56], including the following typical example by Saat et al. [57]. When Japanese (temperate environment) and Malaysian (tropical environment) males were monitored during physical exercise in conditions of ~32°C, 72% relative humidity, the sweat rate of the Malaysian subjects was significantly lower than that of the Japanese [57]. Accordingly, the extent of dehydration in these subjects was significantly less than in their Japanese counterparts [57]. This may seem a paradoxical physiological response as it can be argued that tropical natives should have a greater sweating capacity to dissipate sufficient heat in the high ambient temperatures to which they are habitually exposed. However, the results of the numerous studies which counter this argument [53, 54, 58] suggests that the higher core body temperature repeatedly observed in those people indigenous to warm regions [53, 57–60], affords them enhanced 'dry' heat loss through convection, conduction and radiation. This physiological adaptation, which is usually accompanied by increased skin temperature, is beneficial in reducing the water losses associated with sweating. Furthermore, regulating the core body temperature at a higher set-point reduces the total amount of heat that must be dissipated.

Although the physiological mechanisms behind these adaptations remain unconfirmed, the more pronounced oscillations in circadian rhythm reported in populations living in tropical areas are thought to result from climatic cues [60]. Moreover, the lower subcutaneous fat layers and high body surface area: body mass ratios typically observed in warm climate populations is thought to facilitate more efficient use of dry heat loss [19]. Interestingly, Bae et al. [61] provide evidence that the suppression of sweating in people living in tropical areas occurs as a result of living in a hot climate for a prolonged period, rather than from genotypic differences. Specifically, when Japanese living permanently in Japan were compared to Japanese who had lived 2 years or longer in the tropics, it was found that the tropical residents exhibited an attenuated sweating response, as typically observed in indigenous people from tropical areas [61]. Moreover, sweat volumes were negatively correlated with the duration of having lived in a tropical climate. These findings suggest that people accustomed to living in temperate regions may acquire a similar sweating economy after moving to, and living in, warm climates for a sustained period [61].

Body Composition and Metabolic Rate
Given the fact that regional fat distribution and muscularity are known to vary with ethnicity [62, 63], many would expect those people indigenous to cold climates to have evolved to have a thick, insulating layer of subcutaneous fat to offer optimal protection against the cold elements. Indeed, it has been reported that exposure to even mild cold elicited an increase in heat production three times greater in lean subjects than in their overweight counterparts [64]. Although this energy-efficient, 'blunted' cold response in larger individuals may appear beneficial, there is little evidence to suggest that those people indigenous to cold climates

were able to accumulate particularly high levels of subcutaneous fat. To the contrary, individuals living in cold climates often have skinfold thicknesses lower than those of modern westernized populations [65]. Nevertheless, prior to modernization, individuals indigenous to cold climates were typically relatively muscular and the regional distribution of muscle and fat characteristics of these people is believed to maximize insulation [66].

It is likely that the failure to accumulate a thick, protective layer of subcutaneous fat may be largely attributed to the high metabolic cost of thermogenesis when living in a cold environment. Notably, the average basal metabolic rate of an indigenous population is negatively correlated with the mean annual temperature of their environment, even after controlling for differences in body size [17;67]. Americans of African origin, for example, have a 'low' metabolic rate (reviewed by Luke et al. [68]) whereas there is much evidence to show that populations living in polar regions have an elevated metabolic rate after adjustments for factors which may have had an effect [69, 70]. Although non-indigenous males of Siberian communities did exhibit slightly elevated metabolic rates, the rates of indigenous males who had lived in the cold conditions for longer remained markedly greater [70].

As eloquently elucidated by Snodgrass et al. [71], the elevated basal metabolic rate of populations living in polar regions is believed to involve a combination of short-term acclimation and genetic adaptation. Although there have been previous reports of seasonal variation in basal metabolic rate, e.g. in Japan, indigenous Siberians are distinct in that metabolic rate remains elevated in the warmer months [71]. This indicates long-term genetic adaptation to the severe cold stress of the polar environment, present in addition to a further, short-term, thyroxin-mediated up-regulation of metabolic rate during the winter months. Thyroid activity is modulated by climatic influences including temperature and day length. Also the thyroid hormones, T_3 (tri-iodo thyronine) and T_4 (thyroxin) are known to promote oxidative metabolism and increase heat production through the uncoupling of oxidative phosphorylation in the mitochondria [72]. Recently, it has been reported that the mitochondrial DNA lineages common in central Asia and Siberia are associated with greater uncoupling and increased metabolic heat production. Moreover, it is hypothesized that natural selection may have favoured gene mutations facilitating this increased heat production, ultimately making colonization of cold, northern regions possible (reviewed by Leonard et al. [70]).

Peripheral Vasoconstriction
Peripheral vasoconstriction is an invaluable physiological mechanism for conserving heat in cold conditions. Reduced perfusion allows superficial tissues to form an insulating shell that protects against heat loss by reducing the temperature gradient between the skin and the environment [66]. However, the associated

shunting of blood to the body core adversely affects cardiac function and may also negatively affect the use of the hands and feet [66, 71]. Although the underlying mechanisms require further study, there is some evidence that prolonged residence in a cold climate has stimulated physiological adaptations which reduce the risk of these potential hazards. Firstly, populations living in cold conditions may be less susceptible to cold-induced increases in blood pressure and cardiovascular events than populations living in warmer conditions. Secondly, the Inuit are better able to maintain the temperature of their extremities in response to localized cold stress, in comparison to Oriental and Caucasian subjects from warmer climates [73]. However, the contribution of an underlying genetic adaptation cannot be isolated from that of protective clothing [74]. Although the relative contribution of genetic factors and prior cold-habituation remain unclear, it would appear that populations habituated to a cold climate have at least superior functionality in the cold, if not superior thermoregulation.

Conclusion

There has been extensive study of the interaction between ethnic groups and the varied environmental conditions which they inhabit. While differences have been identified in many of the physiological traits associated with reducing thermal stress, most findings point towards phenotypic rather than genotypic variation. These phenotypic differences occur in response to the particular climate and lifestyle accompanying different environmental conditions. Notably, phenotypic differences display varying degrees of plasticity in response to changing environmental conditions. Sweat rate, for example, continues to adapt within an individual's lifetime after exposure to habitually hot conditions whereas phenotypic changes in a population's skin pigmentation, skeletal morphology or body composition may take hundreds or thousands of years to occur.

The cross-sectional studies, which have produced much of the information in this area, are rather limited because it is difficult to separate the environmental or lifestyle adaptations from genotypic differences. Consider the discovery of about 50 genes, the expression of which are altered depending on their exposure to heat stress, and similarly about 20 genes that are affected by cold [75]; that there are differences in thermoregulatory capacity which have evolved through gene X environment interaction is beyond doubt! However, for future studies to make an impact in identifying genotypic differences, the phenotypical changes arising from diversity in environmental conditions will need to be controlled before there is any intervention (exposure to hot or cold conditions). Whilst this imposes logistical difficulties, people from different populations are spread sufficiently around the planet to make this study design a reality.

References

1 Kottek M, Grieser J, Beck C, Rudolf B, Rubel F: World map of the Köppen-Geiger climate classification updated. Meteorol Z 2006;15:259–263.
2 Gowlett JA: Archaeology: out in the cold. Nature 2001;413:33–34.
3 Pavlov P, Svendesen JI, Indrelid S: Human presence in the European Arctic nearly 40,000 years ago. Nature 2001;413:64–67.
4 Ray D: Pre and post Ice Age environmental factors and modern man. J Environ Sociobiol 2007;4:111–115.
5 Ray N, Currat M, Berthier P, Excoffier L: Recovering the geographic origin of early modern humans by realistic and spatially explicit simulations. Genome Res 2005;15:1161–1167.
6 Bogin B, Rios L: Rapid morphological change in living humans: implications for modern human origins. Comp Biochem Physiol [A] 2003;136:71–84.
7 Bergmann C: Ueber die Verhältnisse der Wärmeökonomie der Thiere zu ihrer Grösse. Gottinger Studien 1847;3:595–708.
8 Allen JC: The influence of physical conditions on the genesis of species. Radical Rev 1877;1:108–140.
9 Dietz WH, Marino B, Peacock NR, Bailey RC: Nutritional status of Efe pygmies and Lese horticulturists. Am J Phys Anthropol 1989;78:509–518.
10 Fredriks AM, van Buuren S, Burgmeijer RJ, Meulmeester JF, Beuker RJ, Brugman E, et al: Continuing positive secular growth change in The Netherlands 1955–1997. Pediatr Res 2000;47:316–323.
11 Wells JCK, Cole TJ: Birth weight and environmental heat load: a between-population analysis. Am J Phys Anthropol 2002;119:276–282.
12 Foster MW, Sharp RR: Race, ethnicity, and genomics: social classifications as proxies of biological heterogeneity. Genome Res 2002;12:844–850.
13 Ncayiyana DJ: Racial profiling in medical research: what are we measuring? S Afr Med J 2007;97:1225–1226.
14 Buskirk ER, Iampietro PF, Bass DE: Work performance after dehydration: effects of physical conditioning and heat acclimatization. J Appl Physiol 1958;12:189–194.
15 Adolph EE: Physiology of Man in the Desert. New York, Interscience Publishers, 1947.
16 Nielsen B, Hales JR, Strange S, Christensen NJ, Warberg J, Saltin B: Human circulatory and thermoregulatory adaptations with heat acclimation and exercise in a hot, dry environment. J Physiol (Lond) 1993;460:467–485.
17 Senay LC, Mitchell D, Wyndham CH: Acclimatization in a hot, humid environment: body fluid adjustments. J Appl Physiol 1976;40:786–796.
18 Bass DE, Buskirk ER, Iampietro PF, Mager M: Comparison of blood volume during physical conditioning, heat acclimatization and sedentary living. J Appl Physiol 1958;12:186–188.
19 Senay LC Jr: Effects of exercise in the heat on body fluid distribution. Med Sci Sports 1979;11:42–48.
20 Senay LC Jr: Changes in plasma volume and protein content during exposures of working men to various temperatures before and after acclimatization to heat: separation of the roles of cutaneous and skeletal muscle circulation. J Physiol (Lond) 1972;224:61–81.
21 Adams WC, Fox RH, Fry AJ, MacDonald IC: Thermoregulation during marathon running in cool, moderate, and hot environments. J Appl Physiol 1975;38:1030–1037.
22 Armstrong LE, Maresh CM: The induction and decay of heat acclimatisation in trained athletes. Sports Med 1991;12:302–312.
23 Rintamaki H: Human responses to cold. Alaska Med 2007;49:29–31.
24 Golden FS, Tipton MJ: Human adaptation to repeated cold immersions. J Physiol (Lond) 1988;396:349–363.
25 Shephard RJ: Metabolic adaptations to exercise in the cold: an update. Sports Med 1993;16:266–289.
26 Shephard RJ: Adaptation to exercise in the cold. Sports Med 1985;2:59–71.
27 Taylor NAS: Ethnic differences in thermoregulation: genotypic versus phenotypic heat adaptation. J Therm Biol 2006;31:90–104.
28 Roberts DF: Body weight, race and climate. Am J Phys Anthropol 1953;11:533–558.
29 Ruff CB: Climatic adaptation and hominid evolution: the thermoregulatory imperative. Evolut Anthropol. Issues, news, rev 1993;2.
30 Houghton P: People of the Great Ocean: Aspects of Human Biology of the Early Pacific. New York, Cambridge University Press, 1996.
31 Lum JK, Jorde LB, Schiefenhovel W: Affinities amongst Melanesians, Micronesians, and Polynesians: a neutral biparental genetic perspective. Hum Biol 2002;74:413–430.
32 Hernandez M, Fox CL, Garcia-Moro C: Fueguian cranial morphology: the adaptation to a cold, harsh environment. Am J Phys Anthropol 1997;103:103–117.
33 Irmak MK, Ozcan O: Human diversity, environmental adaptation and neural crest. Med Hypotheses 1997;48:407–410.

34 Harrison GA, Tanner JM, Pilbeam DR, Baker PT: Human Biology. Oxford, Oxford University Press, 1988, pp 305–507.

35 Irmak MK, Korkmaz A, Erogul O: Selective brain cooling seems to be a mechanism leading to human craniofacial diversity observed in different geographical regions. Med Hypotheses 2004;63: 974–979.

36 Franciscus RG, Long JC: Variation in human nasal height and breadth. Am J Phys Anthropol 1991; 85:419–427.

37 Johnson EF, McClure J, Herron P, Baskerville KA: Anatomical variaton, human diversity, and environmental adaptation. J Natl Med Assoc 1993;85: 337–338.

38 Kollias N, Sayre RM, Zeise L, Chedekel MR: Photoprotection by melanin. J Photochem Photobiol [B] 1991;9:135–160.

39 Pandolf KB, Grange RW, Latzka WA, Bland IH, Kraning KK, Gonzalez RR: Human thermoregulatory responses during heat exposure after artificially induced sunburn. Am J Physiol Regul Integr Comp Physiol 1992;262:R610–R616.

40 Daniels F Jr: In Dill DB, Adolph EF (eds): Adaptation to the Environment. Handbook of Physiology. Washington, American Physiological Society, 1964, pp 969–985.

41 Lapunzina P: Ultraviolet light-related neural tube defects? Am J Med Genet 1996;67:106.

42 Omaye ST: Nutrient deficiencies and pregnancy outcome; in Sharma RP (ed): Dietary Factors and Birth Defects. San Francisco, Pacific Division AAAS, 1993, pp 12–41.

43 Cosentino MJ, Pakyz RE, Fried J: Pyrimethamine: an approach to the development of a male contraceptive. Proc Natl Acad Sci USA 1990;87:1431–1435.

44 Murray FG: Pigmentation, sunlight and nutritional disease. Am Anthropol 1934;36:438–445.

45 Loomis WF: Skin-pigment regulation of vitamin-D biosynthesis in man. Science 1967;157:501–506.

46 Jablonski NG, Chaplin G: The evolution of human skin colouration. J Hum Evol 2000;39:57–106.

47 Chaplin G, Jablonski NG: Hemispheric differences in human skin colour. Am J Phys Anthropol 1998; 107:221–224.

48 Holick MF, MacLaughlin JA, Doppelt SH: Regulation of cutaneous previtamin D$_3$ photosynthesis in man: skin pigment is not an essential regulator. Science 1981;211:590–593.

49 Clemens TL, Henderson SL, Adams JS, Holick MF: Increased skin pigment reduces the capacity of skin to synthesize vitamin D$_3$. Lancet 1982;319:74–76.

50 Scholander PF, Hammel HT, Hart JS, LeMessurier DH, Steen J: Cold adaptation in Australian Aborigines. J Appl Physiol 1958;13:211–218.

51 Scholander PF, Andersen KL, Krog J, Vogt Lorentzen F, Steen J: Critical temperature in Lapps. J Appl Physiol 1957;10:231–234.

52 Visser EP, Dias GJ: A case for reduced skin sensation in high latitude prehistoric Polynesians. Ann Hum Biol 1999;26:131–140.

53 Nguyen M, Tokura H: Sweating and tympanic temperature during warm water immersion compared between Vietnamese and Japanese living in Hanoi. J Hum Ergol 2003;32:9–16.

54 Matsumoto T, Kosaka M, Yamauchi M, et al: Study on mechanisms of heat acclimatization due to thermal sweating: comparison of heat-tolerance between Japanese and Thai subjects. Trop Med 1993;35:23–34.

55 Hori S, Ishizuka H, Nakamura M: Studies on physiological responses of residents in Okinawa to a hot environment. Jpn J Physiol 1976;26:235–244.

56 Lee JB, Matsumoto T, Othman T, Kosaka M: Suppression of the sweat gland sensitivity to acetylcholine applied iontophoretically in tropical Africans compared to temperate Japanese. Trop Med 1997;39: 111–121.

57 Saat M, Tochihara Y, Hashiguchi N, Sirisinghe RG, Fujita M, Chou CM: Effects of exercise in the heat on thermoregulation of Japanese and Malaysian males. J Physiol Anthropol Appl Hum Sci 2005;24: 267–275.

58 Nguyen M, Rutkowska D, Tokura H: Field studies on circadian rhythms of core temperature in tropical inhabitants compared with those in European inhabitants. Biol Rhythm Res 2001;32:547–556.

59 Adam JM, Ferres HM: Observation on oral an rectal temperature in humid tropics and in a temperate climate. J Physiol (Lond) 1954;125:21.

60 Nguyen M, Tokura H: Observations on normal body temperature in Vietnamese and Japanese in Vietnam. J Physiol Anthropol Appl Hum Sci 2002; 21:59–65.

61 Bae JS, Lee JB, Matsumoto T, Othman T, Min YK, Yang HM: Prolonged residence of temperate natives in the tropics produces a suppression of sweating. Pflügers Arch 2006;453:67–72.

62 Rush EC, Goedecke JH, Jennings C, et al: BMI, fat and muscle differences in urban women of five ethnicities from two countries. Int J Obes (Lond) 2007;31:1232–1239.

63 Deurenberg-Yap M, Schmidt G, van Staveren WA, Deurenberg P: The paradox of low body mass index and high body fat percentage amongst Chinese, Malays and Indians in Singapore. Int J Obes Relat Metab Disord 2000;24:1011–1017.

64 Claessens-van Ooijen AMJ, Westerterp KR, Wouters L, Schoffelen PFM, van Steenhoven AA, van Marker Lichtenbelt WD: Heat production and body temperature during cooling and rewarming in overweight and lean men. Obesity 2006;14: 1914–1920.

65 Elsner RW: Skinfold thickness in primitive peoples native to cold climates. Ann NY Acad Sci 1963; 110:503–514.

66 Beall CM, Steegmann ATJ: Human adaptation to climate: temperature, ultra-violet radiation, and altitude; in Stinson S, Bogin B, Huss-Ashmore R, O'Rouke D (eds): Human Biology: An Evolutionary and Biocultural Perspective. New York, Wiley-Liss, 2000, pp 163–224.

67 Roberts DF: Basal metabolism, race, and climate. J R Anthropol Inst GB Ire 1952;82:169–183.

68 Luke A, Dugas L, Kramer H: Ethnicity, energy expenditure and obesity: are the observed black/white differences meaningful? Curr Opin Endocrinol Diabetes Obes 2007;14:370–373.

69 Snodgrass JJ, Leonard WR, Tarskaia LA, Alekseev VP, Krivoshapkin VG: Basal metabolic rate in the Yakut (Sakha) of Siberia. Am J Hum Biol 2005; 17:155–172.

70 Leonard WR, Sorensen MV, Galloway VA, et al: Climatic influences of basal metabolic rates among circumpolar populations. Am J Hum Biol 2002;14: 609–620.

71 Snodgrass JJ, Sorensen MV, Tarskaia LA, Leonard WR: Adaptive dimensions of health research among indigenous Siberians. Am J Hum Biol 2007; 19:165–180.

72 Smals AGH, Ross HA, Kloppenberg PWC: Seasonal variation in serum T3 and T4 levels in man. J Clin Endocrinol Metab 1977;44:998–1001.

73 Livingston SD, Grayson J, Reed LD, Gordon D: Effect of a local cold stress on peripheral temperatures of Inuit, Oriental and Caucasian subjects. Can J Physiol Pharmacol 1978;56:877–881.

74 Donaldson GC, Ermakov SP, Komarov YM, McDonald CP, Keatinge WR: Cold related mortalities and protection against the cold in Yakutsk, eastern Siberia: observation and interview study. Br Med J 1998;317:978–982.

75 Sonna LA, Fujita J, Gaffin SL, Lilly CM: Effects of heat and cold stress on mammalian gene expression. J Appl Physiol 2002;92:1725–1742.

Michael I. Lambert, PhD
MRC/UCT Research Unit for Exercise Science and Sports Medicine
PO Box 115
Newlands, 7725 (South Africa)
Tel. +27 21 650 4558, Fax +27 21 686 7530, E-Mail Mike.Lambert@uct.ac.za

Lambert · Mann · Dugas

Marino FE (ed): Thermoregulation and Human Performance. Physiological and Biological Aspects.
Med Sport Sci. Basel, Karger, 2008, vol 53, pp 121–129

Exercise Heat Stress and Metabolism

Toby Mündel

Institute of Food, Nutrition and Human Health, Massey University,
Palmerston North, New Zealand

Abstract

Apart from few studies, the majority of the research conducted on the effects of heat stress on energy metabolism during exercise has only been done so in the past two decades. Whilst increasing exercise duration under conditions of heat stress favours the oxidation of carbohydrate (CHO) and appears to increase the rate of muscle glycogenolysis, total CHO oxidation is often less and levels of muscle glycogen remain much higher at the point of fatigue when compared with the same exercise without heat stress. Furthermore, supplementing CHO during exercise in the heat appears to exert an ergogenic effect that is not related to 'peripheral' but rather 'central' factors. However, there may be a role for the excess ammonia (NH_3) produced in the exercising muscle during heat stress, as cerebral uptake and subsequent metabolism of NH_3 may have detrimental effects on cerebral function. Recent exciting results point toward an increased cerebral CHO uptake relative to that of O_2, termed the cerebral metabolic ratio (CMR) during exercise with heat stress, although a causative link between this and reduced exercise performance has yet to be identified. Therefore, it appears that despite a shift towards greater CHO utilisation in both skeletal muscular and cerebral metabolism, these responses have ultimately not proved limiting to exercise with heat stress.

Despite an apparent lack of real research interest and published evidence until the ~1990s, the common view held by exercise scientists has been that the addition of heat stress to exercise provokes greater changes in metabolism when compared to similar exercise in more temperate climates. In spite of this view, it is interesting to note that Galen [1] added his suggestion of temperature to Aristotle's [2] doctrine of the five senses – sight, hearing, smell, taste and touch – some 1,600 years earlier than the addition of muscular senses by Erasmus Darwin [3]; an early indication perhaps of the hierarchical nature with which our body's homeostatic mechanisms operate?

Prolonged exercise itself often causes a marked rise in body temperature, that is a rise in core (T_{core}) and muscle (T_{mus}) temperatures that is thought to have a beneficial effect by increasing the rate of metabolic reactions, such as actin/myosin interaction and glycolytic and oxidative pathways within the body (Q_{10} effect). However, the additional heat stress placed upon the body whilst exercising for a prolonged period in hot conditions exacerbates this rise in body temperature and it is now well established that this heat stress reduces rather than improves both the capacity [4] and performance [5] of endurance exercise.

The present chapter will discuss possible alterations in metabolism and the site of such an occurrence, and whether these purported changes become limiting during prolonged exercise with heat stress.

Skeletal Muscle Metabolism during Exercise and Heat Stress

At intensities between 60 and 90% maximum O_2 uptake ($\dot{V}O_{2\,max}$) muscle glycogen stores and levels of blood glucose are generally thought to be the limiting factors to exercise endurance in moderate ambient conditions, with depletion of these stores or hypoglycaemia being associated with earlier fatigue or a decreased performance. There have been consistent observations of an increased respiratory exchange ratio (RER), reflective of a greater carbohydrate (CHO) and reduced fat oxidation, during exercise in the heat when compared to exercise at cooler ambient temperatures [4, 6], suggesting that glycogen stores may become depleted sooner when exercising in the heat. However, not all studies report an increased RER [7] and total CHO oxidation is often lower as fatigue is reached at an earlier time-point during exercise in the heat when compared to cooler conditions [4]. However, estimates of whole-body substrate oxidation may not reflect actual substrate utilization in the exercising muscle. For example, an increased muscle CHO oxidation could be exactly balanced by a decreased consumption in some other tissue, although since the consumption of CHO by muscle far exceeds that of other sources, it would be difficult to find another tissue that would balance the muscle utilisation.

In a series of studies [6–11], it was demonstrated that a greater rate of muscle glycogenolysis occurred during exercise in the heat (40°C) than in the cool (20°C). During 40 min of cycling at 70% $\dot{V}O_{2\,max}$ an increased muscle lactate and glycogenolysis was observed at 40°C when compared to the same protocol at 20°C [6]. Using the same exercise protocol but following 7 days of heat acclimation, Febbraio et al. [8] also demonstrated that muscle glycogenolysis and lactate production were attenuated after heat acclimation. These changes were accompanied by a reduction in T_{core}, T_{mus} and catecholamine response, and so in a subsequent study [9] the rise in T_{core} was blunted by cycling for 40 min at 70% $\dot{V}O_{2\,max}$ in 3°C as opposed 20°C. The authors found a reduced muscle glycogenolysis was accompanied by a lower T_{mus} and catecholamine response. Studies in which direct infusion of adrenaline to mimic levels observed during exercise at 40°C [10] and water-perfused cuffs used to manipulate T_{mus} independently of T_{core} [11], were able to identify the mechanisms of an increased muscle lactate production, glycogenolysis and CHO oxidation during exercise in the heat (40°C) as opposed to the cool (20°C), being due in part but possibly not limited to a higher T_{mus} and sympatho-adrenal response secondary to the greater rise in T_{core}.

Therefore, from the available evidence, it can be concluded that during exercise with heat stress there is a greater reliance on CHO as a substrate and that the greater depletion of muscle glycogen stores may limit exercise, although it is worth noting that both greater reliance and substrate oxidation occur as a consequence of a larger rise in T_{core} when compared to exercise in more temperate environments.

The obvious question, then, is whether the greater substrate depletion during exercise heat stress causes the reduced exercise capacity [4] and performance [5]?

If it is an increased CHO oxidation and muscle glycogenolysis that limits exercise in hot conditions, then it should be possible to prolong or improve such exercise through exogenous feeding of CHO. Carter et al. [12] sought to answer this question by exercising subjects at 35°C in either compensable (60% $\dot{V}O_{2\,max}$) conditions so that thermoregulatory capacity permitted CHO reserves to be depleted or in uncompensable (73% $\dot{V}O_{2\,max}$) conditions. Subjects received either a solution containing 6.4% CHO or coloured placebo with the results surprisingly showing time-to-exhaustion being extended with CHO supplementation in both compensable and uncompensable heat stress. Important to note was that CHO oxidation rates remained high and similar between CHO and placebo trials coupled with no apparent signs of hypoglycaemia in any trial. This led the authors to conclude that in the absence of any (peripheral) metabolic explanation, the central nervous system was a more likely target for these ergogenic effects. The results of this study support previous observations in runners [13].

Perhaps the most definitive study to answer whether it is substrate depletion that limits exercise in the heat again comes from the Melbourne laboratory [7]. Endurance-trained subjects cycled to exhaustion at 70% $\dot{V}O_{2\,max}$ in three environmental conditions: 3, 20 and 40°C. Exercise time was shortest during 40°C and whilst muscle glycogen levels at fatigue had decreased from resting values in all three trials, they were significantly higher at 40°C and remained above $300\,mmol \cdot kg^{-1}$ dry weight (fig. 1) leading the authors to conclude that 'fatigue during exercise in the heat is related to processes other than CHO availability' [7, p. 904].

Ammonia (NH$_3$) – Link between the Exercising Muscle and Brain?

The data from the study mentioned previously [7] and others [14, 15] do suggest that other metabolic pathways may be involved in the fatigue process during exercise in the heat. Marino et al. [14] had subjects complete 30 min of treadmill running at 70% peak running speed followed by a self-paced 8-km performance run; trials were conducted in cool (15°C) and hot (35°C) humid conditions. Levels of

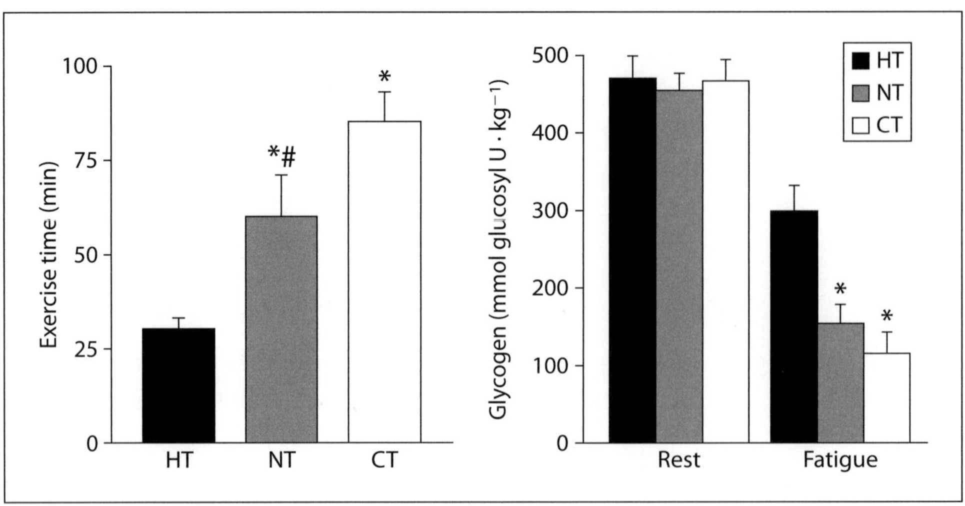

Fig. 1. Exercise time to exhaustion and muscle glycogen before (Rest) and after (Fatigue) cycling at 70% O$_2$ uptake in 40°C (HT), 20°C (NT) and 3°C (CT). *Different from HT, p < 0.05. #Different from CT, p < 0.05. From Parkin et al. [7], with permission.

plasma NH$_3$ remained constant and were similar between trials during the 30 min of steady-state running; however, thereafter plasma NH$_3$ increased to a greater extent during the self-paced performance run in 35°C compared to 15°C. The results of this study differ slightly from those of Snow and colleagues [15] who cycled subjects for 40 min at 70% $\dot{V}O_{2\,max}$ in 20 and 40°C. Both plasma and muscle concentrations of NH$_3$ were higher by the end of exercise in the hot compared to the cool trial. The difference in results between these studies is most likely due to the mode of exercise, as it has been shown that cycling produces a greater NH$_3$ accumulation than does running [16]. Nevertheless, the source of this additional NH$_3$ during exercise in the heat appears to be due to degradation of muscle amino acids rather than the deamination of AMP to IMP + NH$_3$ by purine catabolism [6, 14], although it has been suggested that mitochondrial disruption could account for the increased concentrations of IMP and NH$_3$ following exercise with heat stress [7]. Regardless, it is unlikely that contribution of protein to total energy turnover is anything other than minimal.

However, the potential consequences of this additional NH$_3$ production in and subsequent release by the exercising muscle in the heat have been proposed to have an auxiliary effect on cerebral factors [see 17, for review]. A greater release of NH$_3$ by the exercising muscle and subsequent delivery to the brain, as evidenced by higher systemic levels could lead to an excessive intra- and extra-cellular cerebral accumulation as NH$_3$ is readily able to cross the blood-brain barrier; this may

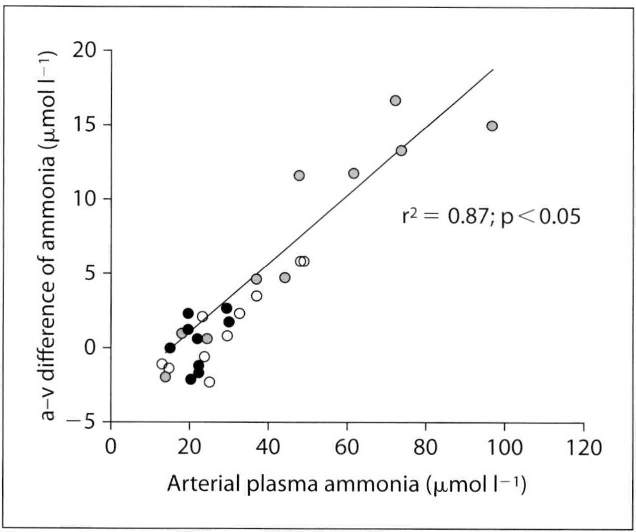

Fig. 2. Cerebral arterial-internal jugular venous (A-jV) differences of NH_3 versus arterial plasma NH_3 at rest (black symbols), after 30 min (open symbols) and 180 min (red/grey symbols) of exercise. Symbols represent individual values. From Nybo et al. [18], with permission.

result in disruption of cerebral neurotransmitter homeostasis with possible detrimental effects on cerebral function [17]. The brain has no effective uric cycle and therefore the removal of any excess NH_3 is dependent on the synthesis of glutamine from glutamate and NH_3. Glutamate is the precursor for gamma-aminobutyric acid (GABA) synthesis; therefore, hyperammonemia could effectively alter cerebral levels of glutamine, glutamate and GABA, with potential detrimental effects on cerebral neurotransmission, circulation and metabolism [17]. Although cerebral NH_3 uptake has been demonstrated in maximal and submaximal exercise (fig. 2) and is thought to be related to perception of effort [18], to the author's knowledge no measures have been made during exercise with and without heat stress; this would seem a logical step to take. Furthermore, an associative relationship but no causality has yet been established between the hyperammonemia observed during exhaustive exercise (such as in the heat) and a reduced exercise capacity/performance.

Cerebral Metabolism during Exercise and Heat Stress

During exercise, skeletal muscle is the most metabolically active tissue and therefore has received a large part of the research attention with regards to changes in metabolism; exercise in the heat is no exception as evidenced by the studies already

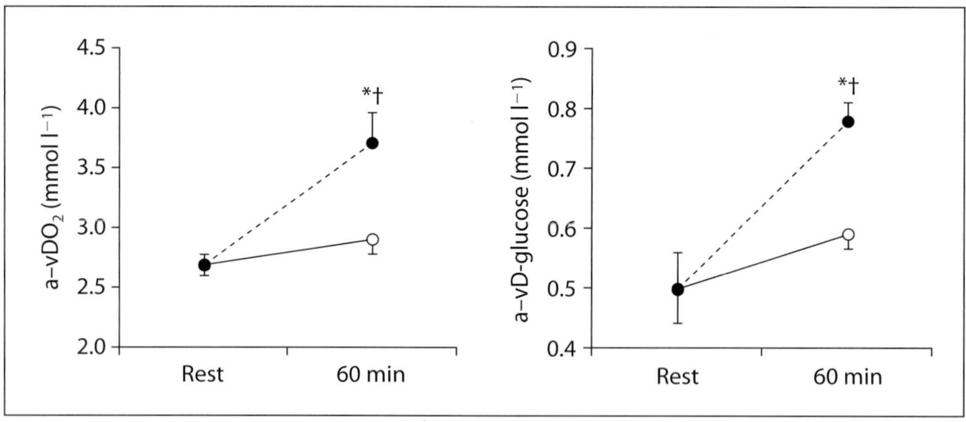

Fig. 3. Cerebral arteriovenous difference (a-vD) of O_2 and glucose at rest and following 60 min of exercise with (closed circles) and without (open circles) heat stress. *Main effect time, $p < 0.05$. †Main effect trial, $p < 0.05$. Re-drawn from Nybo et al. [20], with permission.

discussed. For some time now, the laboratory of Niels Secher in Copenhagen has centred much of its research around 'the exercising brain'. In particular, recent studies were the first to establish an increased energy turnover in the human brain during exercise with, as opposed to without heat stress [19, 20]. Nybo et al. [19, 20] cycled subjects for 65 min at 50% $\dot{V}O_{2\,max}$ in 20°C with (HOT) and without (CON) insulative clothing whilst measuring the arterial-internal jugular venous (A-jV) differences for O_2 and glucose. Exercise in HOT resulted in a 1.6°C higher T_{core} and a significantly greater A-jV difference in cerebral O_2 and glucose (fig. 3).

The brain relies on aerobic metabolism, with a preference for a constant systemic glucose supply as glycogen stores in neuronal tissue are limited. The cerebral metabolic ratio (CMR) – defined as the ratio of O_2 uptake to that of CHO and is normally 6 at rest, reflecting the metabolism of six carbon atoms in glucose – is expressed independent to any changes in cerebral blood flow that may occur during exercise heat stress [see 21, for review]. There is also the capacity for the brain to metabolise other substrates such as lactate; glycolysis in astrocytes and an uptake from systemic concentrations have been observed, and thus lactate is integrated into the equation for CMR (O_2 uptake/uptake of glucose + 1/2 uptake of lactate) [21]. However, during exercise heat stress neither cerebral uptake nor arterial concentrations of lactate increase and therefore most probably do not contribute to the ~50% decrease in CMR observed [19, 20]. Exercising in the heat is often accompanied by a greater 'mental effort' than when performed in the cool, as demonstrated by higher ratings of perceived exertion (RPE) [17]. Equally, whilst CMR remains stable during prolonged exercise in the cool, once subjects

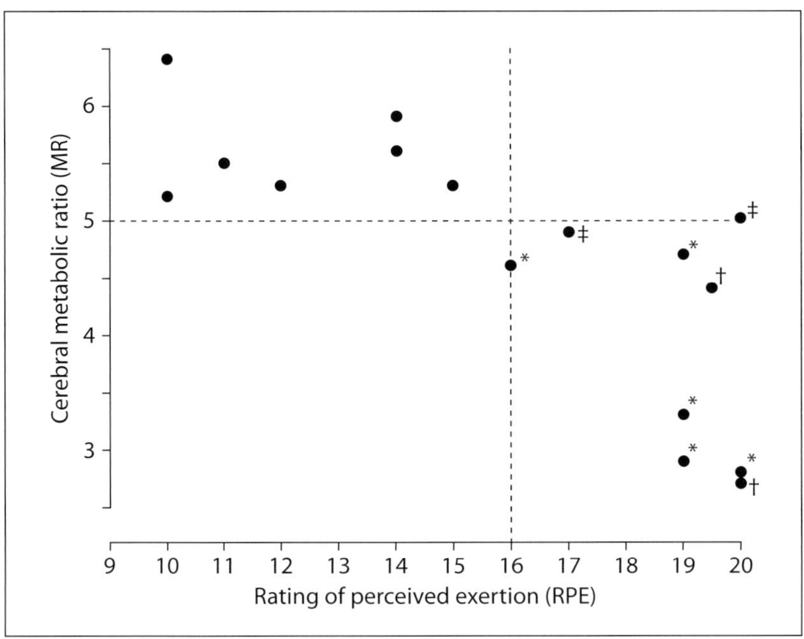

Fig. 4. Cerebral metabolic ratio vs. perceived exertion. The dotted lines represent an apparent threshold beyond which exercise significantly reduces CMR, i.e. an RPE >15. From Dalsgaard [21], with permission.

begin to struggle (RPE >15) CMR decreases, particularly so during exercise and heat stress (fig. 4).

The reduction in CMR is currently poorly understood. Various neuro-humoral factors have been proposed to affect CMR including, but not exclusively, insulin and insulin-like growth factor-1, cortisol, interleukin-6, NH_3 and noradrenalin [21]. Just as previously stated with NH_3 as yet no causality has been confirmed between the decrease in CMR and reduced exercise capacity/performance in the heat, although of interest is the fact that CMR reaches its nadir at fatigue, suggestive of a link between the limit(s) of exercise and cerebral metabolism [21]. However, thus far studies investigating heat stress and CMR have employed open-loop models of exercise (fixed-intensity) with fatigue being the conscious termination of exercise. The ecological validity of such a design is poor, and, as has been argued by Tucker [this vol., pp. 26–38] does not allow the detection of small self-selected changes by an athlete during the exercise, as is the case with a closed-loop model (variable-intensity). Therefore, it seems plausible that future studies measuring CMR are repeated with a closed-loop exercise model or even the RPE clamp design [22] to further ascertain whether changes seen in cerebral metabolism with exercise heat stress are causal or coincidental.

Conclusion

Although traditionally increased body temperature was thought to be of benefit to performance (e.g. warming-up), the addition of heat stress to exercise and resultant hyperthermia is now seen to be a limiting factor. Alterations in skeletal muscle metabolism, such as a greater glycogenolysis and CHO oxidation cannot account for a reduced exercise capacity in the heat. A greater release from muscle and accumulation/metabolism of NH_3 in the brain have been postulated to reduce cerebral function, but further studies to ascertain its role in the aetiology of fatigue in the heat need to be conducted. Recent findings suggest the cerebral CHO uptake is much larger than can be accounted for by the uptake of O_2 when exercising under conditions of heat stress; however, causative effects of fatigue have not yet been established.

References

1 Galen (circa 160 AD): On the Natural Faculties (English translation by Brock, AJ), Book I, Chapter III. London, Heinemann, 1916.
2 Aristotle (circa 350 BC.): Aristotle's Psychology: A Treatise on the Principles of Life (English translation by Hammond, WA). New York, MacMillan, 1902.
3 Darwin E: Zoonomia, or The Laws of Organic Life. London, Johnson, 1794.
4 Galloway SDR, Maughan RJ: Effects of ambient temperature on the capacity to perform prolonged cycle exercise in man. Med Sci Sports Exerc 1997;29: 1240–1249.
5 Tatterson AJ, Hahn AG, Martin DT, Febbraio MA: Effects of heat stress on physiological responses and exercise performance in elite cyclists. J Sci Med Sport 2000;3:186–193.
6 Febbraio MA, Snow RJ, Stathis CG, Hargreaves M, Carey MF: Effect of heat stress on muscle energy metabolism during exercise. J Appl Physiol 1994;77: 2827–2831.
7 Parkin JM, Carey MF, Zhao S, Febbraio MA: Effect of ambient temperature on human skeletal muscle metabolism during fatiguing submaximal exercise. J Appl Physiol 1999;86:902–908.
8 Febbraio MA, Snow RJ, Hargreaves M, Stathis CG, Martin IK, Carey MF: Muscle metabolism during exercise and heat stress in trained men: effect of acclimation. J Appl Physiol 1994;76:589–597.
9 Febbraio MA, Snow RJ, Stathis CG, Hargreaves M, Carey MF: Blunting the rise in body temperature reduces muscle glycogenolysis during exercise in humans. Exp Physiol 1996;81:685–693.
10 Febbraio MA, Lambert DL, Starkie RL, Proietto J, Hargreaves M: Effect of epinephrine on muscle glycogenolysis during exercise in trained men. J Appl Physiol 1998;84:465–470.
11 Starkie RL, Hargreaves M, Lambert DL, Proietto J, Febbraio MA: Effect of temperature on muscle metabolism during submaximal exercise in humans. Exp Physiol 1999;84:775–784.
12 Carter J, Jeukendrup AE, Mundel T, Jones DA: Carbohydrate supplementation improves moderate and high-intensity exercise in the heat. Pflügers Arch 2003;446:211–219.
13 Millard-Stafford ML, Sparling PB, Rosskopf LB, Dicarlo LJ: Carbohydrate-electrolyte replacement improves distance running performance in the heat. Med Sci Sports Exerc 1992;8:934–940.
14 Marino FE, Mbambo Z, Kortekaas E, Wilson G, Lambert MI, Noakes TD, Dennis SC: Influence of ambient temperature on plasma ammonia and lactate accumulation during prolonged submaximal and self-paced running. Eur J Appl Physiol 2001; 86:71–78.
15 Snow RJ, Febbraio MA, Carey MF, Hargreaves M: Heat stress increases ammonia accumulation during exercise in humans. Exp Physiol 1993;78: 847–850.
16 Bouckaert J, Pannier JL: Blood ammonia response to treadmill and bicycle exercise in man. Int J Sports Med 1995;16:141–144.
17 Nybo L, Secher NH: Cerebral perturbations provoked by prolonged exercise. Prog Neurobiol 2004; 72:223–261.

18 Nybo L, Dalsgaard MK, Steensberg A, Moller K, Secher NH: Cerebral ammonia uptake and accumulation during prolonged exercise in humans. J Physiol (Lond) 2005;563:285–290.

19 Nybo L, Moller K, Volianitis S, Nielsen B, Secher NH: Effects of hyperthermia on cerebral blood flow and metabolism during prolonged exercise in humans. J Appl Physiol 2002;93:58–64.

20 Nybo L, Nielsen B, Blomstrand E, Moller K, Secher N: Neurohumoral responses during prolonged exercise in humans. J Appl Physiol 2003;95:1125–1131.

21 Dalsgaard MK: Fuelling cerebral activity in exercising man. J Cereb Blood Flow Metab 2006;26:731–750.

22 Tucker R, Marle T, Lambert EV, Noakes TD: The rate of heat storage mediates an anticipatory reduction in exercise intensity during cycling at a fixed rating of perceived exertion. J Physiol (Lond) 2006; 574:905–915.

Toby Mündel, PhD
Institute of Food, Nutrition and Human Health, Massey University
Private Bag 11 222
Palmerston North, 4471 (New Zealand)
Tel. +64 6 350 5799, Fax +64 6 350 5657, E-Mail T.Mundel@massey.ac.nz

Author Index

Subject Index